# SpringerBriefs in Electrical and Computer Engineering

For further volumes:
http://www.springer.com/series/10059

Touhid Bhuiyan

# Trust for Intelligent Recommendation

 Springer

Touhid Bhuiyan
Dutton Park, QLD
Australia

ISSN  2191-8112          ISSN 2191-8120   (electronic)
ISBN  978-1-4614-6894-3     ISBN 978-1-4614-6895-0   (eBook)
DOI 10.1007/978-1-4614-6895-0
Springer New York Heidelberg Dordrecht London

Library of Congress Control Number: 2013932127

Printed on acid-free paper

Springer is part of Springer Science+Business Media (www.springer.com)

*To*
*Anna and Rohn*

# Preface

Recommender systems are one of the recent inventions to deal with ever growing information overload in relation to the selection of goods and services in a global economy. Collaborative Filtering (CF), is one of the most popular techniques in recommender systems. The CF recommends items to a target user based on the preferences of a set of similar users known as the neighbors, generated from a database made up of the preferences of past users. With sufficient background information of item ratings, its performance is promising enough but research shows that it performs very poorly in a cold start situation where there is not enough previous rating data. As an alternative to ratings, *trust* between the users could be used to choose the neighbor for recommendation making. Better recommendations can be achieved using an inferred trust network which mimics the real world "friend of a friend" recommendations. To extend the boundaries of the neighbor, an effective trust inference technique is required. The research work presented in this book proposes a new trust interference technique called Directed Series Parallel Graph (*DSPG*), which performs better than other popular trust inference algorithms such as *TidalTrust* and *MoleTrust*.

Another problem is that reliable explicit trust data are not always available. In real life, people trust "word of mouth" recommendations made by people with similar interests. This is often assumed in the recommender system. By conducting a survey, it has been shown that interest similarity has a positive relationship with trust and this can be used to generate a trust network for recommendation. In this book, the author also proposed a new method called *SimTrust* for developing trust networks based on user's interest similarity in the absence of explicit trust data. To identify the interest similarity, user's personalized tagging information has used. However, the research focused on what resources the user chooses to tag, rather than the text of the tag applied. The commonalities of the resources being tagged by the users can be used to form the neighbors used in the automated recommender system. The experimental results show that the proposed tag-similarity based method outperforms the traditional collaborative filtering approach which usually uses rating data.

# Acknowledgments

I would like to express my sincere gratitude and deep appreciation to Prof. Yue Xu, Queensland University of Technology, Australia, the project leader, for her guidance, encouragement, support and constructive comments in my research, and for her knowledge and wisdom in Computer Science. The achievements in this book would not possible without her cordial advice and supervision. I am also thankful to Prof. Audun Jøsang, University of Oslo, Norway, for providing instrumental inputs and guidance for my research and for helping me with my publications. Thanks to Prof. Kerry Raymond, for her continuous encouragement to overcome any obstacle of the research. I would also like to thank the entire group of volunteer survey respondents for their valuable time and opinions and the members of my family for their continuous support and understanding throughout the period of research. Finally, special thanks to my faculty, which has provided me the comfortable research environment with needed facilities and financial support including my scholarship and travel allowances over the period of my candidature. I would especially like to thank all the members of our reading seminar group for offering invaluable advice and comments regarding my research work.

Acknowledgments

# Contents

# Author's Other Publications Related to this Book

Bhuiyan, T. (2010). A Survey on the Relationship between Trust and Interest Similarity in Online Social Networks. *Journal of Emerging Technologies in Web Intelligence*, 2(4), 291–299.

Bhuiyan, T. and Josang, A. (2010). Analysing Trust Transitivity and the Effects of Unknown Dependence. International Journal of Engineering Business Management, Vol 2, Issue 1, pp. 23–28.

Bhuiyan, T., Jøsang, A., & Xu, Y. (2010). Managing Trust in Online Social Networks. In B. Furht, (Ed.), *Handbook of Social Network Technologies and Applications.* Springer, USA, 447–496.

Bhuiyan, T., Jøsang, A., & Xu, Y. (2010). Trust and Reputation Management in Web-based Social Network. In Z. Usmani, (Ed.), *Web Intelligence and Intelligent Agents.* In-Tech, Croatia, 207–232.

Bhuiyan, T., Xu, Y., Jøsang, A., Liang, H., & Cox, C. (2010). Developing Trust Networks Based on User Tagging Information for Recommendation Making. The 11th *International Conference on Web Information System Engineering.* 12–14 Dec, 2010, Hong Kong, China.

Bhuiyan, T., Xu, Y., & Jøsang, A. (2010). The Potential Relationship between Trust and Interest Similarity. The 11th *International Conference on Internet Computing,* 12–15 July, 2010, Nevada, USA.

Bhuiyan, T., Xu, Y., & Jøsang, A. (2010). SimTrust: A New Method of Trust Network Generation. The 6th International Symposium on Trusted Computing and Communications, 11–13 Dec, 2010, Hong Kong, China.

Bhuiyan, T., Xu, Y., & Jøsang, A. (2010). A Review of Trust in Online Social Networks to Explore New Research Agenda. The 11th *International Conference on Internet Computing,* 12–15 July, 2010, Nevada, USA.

Bhuiyan, T. (2009). Taxonomy of Online Opinion Mining Research. International Transactions on Computer Science and Engineering, Vol 57, Number 1, pp. 220–230.

Bhuiyan, T., Jøsang, A., & Xu, Y. (2009). An Analysis of Trust Transitivity Taking Base Rate into Account. The Symposia of the 6th International Conference on Ubiquitous Intelligence and Computing, 7–9 July, 2009, Brisbane, Australia.

Bhuiyan, T., Xu, Y., & Jøsang, A. (2009). State-of-the-Art Review on Opinion Mining from Online Customers' Feedback. The 9th Asia-Pacific Complex Systems Conference, 4–7 November, 2009, Tokyo, Japan.

Bhuiyan, T., Xu, Y., & Jøsang, A. (2008). Integrating Trust with Public Reputation in Location-based Social Networks for Recommendation Making. The Workshop of the IEEE/WIC/ACM International Conference on Web Intelligence and Intelligent Agent Technology, Dec 9–12, 2008, Sydney, Australia.

Jøsang, A., Bhuiyan, T., Xu, Y., & Cox, C. (2008). Combining Trust and Reputation Management for Web-Based Services. The 5th International Conference on Trust, Privacy and Security in Digital Business, 1–5 September, 2008, Turin, Italy.

Jøsang, A., & Bhuiyan, T. (2008). Optimal Trust Network Analysis with Subjective Logic. The Second International Conference on Emerging Security Information, Systems and Technologies, August 25–31, 2008. Cap Esterel, France.

# Chapter 1
# Introduction

**Abstract** We are currently experiencing an era of information explosion. A study on "how much information" conducted initially in 2000 and updated in 2003 by the School of Information Management and Systems of the University of California, Berkeley found that we produce more than 5 Exabytes of unique information per year, which is roughly 800 megabytes for each human on earth. An Exabyte is a billion Gigabytes, or $10^{18}$ bytes. Printed documents of all kinds comprise only 0.003 % of the total. As of April 2012, there are more than 140 million registered web domains. It is estimated that around half of them are actively maintained and kept current, representing more than 3,200 million pages of information growing daily! The information explosion attacks us from every angle. For example the Internet Movie Database (IMDb) states that it lists 1,472,014 individual film/TV productions (games and more), and some 3,128,262 names of people who have worked on these productions (McIlroy 2010). An Internet search engine for blogs called *Technorati* http://technorati.com, indicates that it has indexed 133 million blog entries since 2002. These scenarios explained above indicate that the Internet changed the way people deal with information. The incredible growth of information on the Internet is giving more choices but at the same time creating one of the biggest challenges of the Internet, which is the efficient processing of this growing volume of information.

**Keywords** Recommender systems · Trust · Inference · User tag · Web 2.0 · Social networks

We are currently experiencing an era of information explosion. A study on "how much information" conducted initially in 2000 and updated in 2003 by the School of Information Management and Systems of the University of California, Berkeley found that we produce more than 5 Exabytes of unique information per year, which is roughly 800 megabytes for each human on earth. An Exabyte is a billion Gigabytes, or $10^{18}$ bytes. Printed documents of all kinds comprise only 0.003 % of the total. As of April 2012, there are more than 140 million registered web domains. It is estimated that around half of them are actively maintained and kept

T. Bhuiyan, *Trust for Intelligent Recommendation*,
SpringerBriefs in Electrical and Computer Engineering,
DOI: 10.1007/978-1-4614-6895-0_1, © The Author(s) 2013

current, representing more than 3,200 million pages of information growing daily! The information explosion attacks us from every angle. For example the Internet Movie Database (IMDb) states that it lists 1,472,014 individual film/TV productions (games and more), and some 3,128,262 names of people who have worked on these productions (McIlroy 2010). An Internet search engine for blogs called *Technorati* http://technorati.com, indicates that it has indexed 133 million blog entries since 2002. These scenarios explained above indicate that the Internet changed the way people deal with information. The incredible growth of information on the Internet is giving more choices but at the same time creating one of the biggest challenges of the Internet, which is the efficient processing of this growing volume of information.

Recently recommender systems which extract useful information from the Internet and generate recommendations have emerged to help users overcome the exponentially growing information overload problem.

To be able to cope with the magnitude, diversity and dynamic characteristics of web data, people need the assistance of intelligent software agents for searching, sorting and filtering the available information (Etzioni 1996). To achieve this, the field of personalization has grown dramatically in recent years. Personalized recommender systems attempt to penetrate users' diverse demands and generate tailored recommendations. Hagen defines personalization as "the ability to provide content and services tailored to individuals based on knowledge about their preferences and behaviors" (Hagen et al. 1999). Most personalization systems are based on some type of user profile (Mobasher et al. 2002; Shepitsen et al. 2008). A user profile is a collection of information about a user, including demographic information, usage information, and interests or goals either explicitly stated by the user or implicitly derived by the system (Zigoris and Zhang 2006; Arapakis et al. 2009; Liang 2011). An individual user profile describes only that user's personal interests and other information. Explicit item ratings are based on typical user information from which recommender systems generate recommendations to users. However, the available user explicit ratings are usually not sufficient to profile users accurately. There are many different techniques and systems which have already been developed and implemented in different domains for automated recommendations. But most of the existing research on recommender systems focuses on developing techniques to better utilize the available information resources to achieve better recommendation quality. Because the amount of available user profile data remains insufficient; these techniques have achieved only limited improvements to overall recommendation quality (Park et al. 2006; Weng 2008). Therefore, exploring new data sources is desirable for efficient profiling and improved personalization.

User trust information is one useful data source. In recent years, incorporating trust models into recommender systems has attracted the attention of many researchers (Massa and Avesani 2007; Ziegler and Golbeck 2007; Adomavicius and Tuzhilin 2005). Their emphasis is on generating recommendations based upon trusted peers' opinion, instead of the traditional most similar users' opinion. Massa and Avesani (2007) presented evidence that trust-based recommender systems can be more effective than traditional collaborative filtering based systems. They argued that data sparsity causes serious weaknesses in collaborative filtering systems. However, they have assumed that the trust network already exists with the users' explicit trust data. They did not consider a situation where the trust network is not available. Trust-based recommenders can make recommendations as long as a new user is connected to a large enough component of the trust network. Ziegler and Golbeck (2007) have proposed frameworks for analyzing the correlation between interpersonal trust and interest similarity and have suggested a positive interaction between the two. They have assumed that if two people have similar interests, they most likely would trust each other's recommendations. We have conducted an online survey to investigate the relationship between trust and interest similarity. The results of our survey on the relationship between trust and interest similarity in an online environment also strongly support Ziegler's hypothesis. Motivated by these findings, we propose to use users' interest similarity to form the trust network among the users, irrespective of personal relationship; based only on utility.

Trust has various meanings in different contexts. For this book, we define trust as a subjective probability by which an agent (i.e. a person) can have a level of confidence in another agent in a given scope, to make a recommendation. For example, by saying "the trust value of agent $A$ to agent $B$ is 0.7"; we mean that agent $A$ is happy to rely 70 % on agent $B$ to make a recommendation for a product or service in that particular scope. Here scope is very important; because a person may be happy to rely on another person's recommendation about a movie but may not ready to accept their recommendation about his/her career decision. The trust value between two agents is not scope independent. For different scopes, the value could be different between two particular agents.

Some social networks allow users to assign a value to their peers to express how much to trust them. But research shows that people do not like to express explicit trust values for their peers or friends in a network. Even if the trust value is assigned directly, it may also change over time. Hence, the explicit trust value is generally not available in real world applications and even if available, it is not reliable. To address this issue, we consider a situation where the trust value does not exist. We attempt to generate a virtual trust network based on interest similarity of users using their tagging behavior. Our interest similarity based trust network is not the actual directly-user-assigned explicit trust network but it could be a helpful tool to make recommendations in the absence of a real trust network. Based on the outcome of our survey, we have assumed that users' interest similarity has a positive correlation with trust. Therefore, we use the interest similarity value instead of the explicit trust value to build the similarity based trust network.

A traditional similarity value does not have any direction. For example, the similarity value between two agents $A$ and $B$ is 0.4, which means that $A$ is similar to $B$ of the value 0.4 and also $B$ is similar to $A$ of the value 0.4. But the trust value from $A$ to $B$ is 0.4 does not necessarily mean that from $B$ to $A$ will also be 0.4. For this reason, while interest similarity is used as the trust value, a sophisticated technique is used to calculate the interest similarity between two agents. An example could be helpful to describe the basic idea behind this. Say, user *A* and user *B* like 5 and 10 items respectively. If user *B*'s 10 items include all of the 5 items liked by user A, then if we use traditional technique of similarity calculation, the similarity value of *A* to *B* and *B* to *A* will be a same value. But in our proposed similarity calculation technique, *A* will be 100 % similar to *B* because all the items liked by user *A* are also liked by user *B*. However, unlike the traditional similarity calculation, user *B* will be 50 % similar to user *A*. As among the 10 items liked by user *B*, only 5 items are also liked by user *A*. This directed property of similarity is compatible to directed trust. Because, the trust value between two agents is not always equal in each direction. If *A* trust *B* value is 0.3, then *B* trust *A* value could be same or any other value.

In a collaborative filtering recommender system, a set of similar users based on previous common ratings are known as neighbors. Whereas, in a trust network, any node connected directly or indirectly to a source node with a trust value can be used as a trusted neighbor to make recommendations. The trust value from the source node to any of the indirectly connected nodes can be derived by using a trust inference technique.

Better recommendations can be achieved using an inferred trust network which mimics the real world "friend of a friend" recommendations. A web-of-trust between community members is created which is often used in recommender systems helping users of e-commerce applications to get an idea about the trust-worthiness of their mostly-personally-known cooperation partners. However, these web-of-trusts are often too sparse to be helpful in practice. A user may have experience with only a very small fraction of the community members. To extend the boundary of the neighbor, an effective trust inference technique is required. A number of trust inference techniques are available in the current literature, along with their strengths and weaknesses. These algorithms differ from each other in the way they compute trust values and infer those values in the networks. Golbeck's (2005) *TidalTrust* is the most referred trust inference technique which performs a modified Breadth First Search (BFS) in the trust network to compute trust between users who are not directly connected in the trust network. Basically, it finds all users with the shortest path distance from the source user and aggregates their ratings weighted by the trust between the source user and these users. Since *TidalTrust* only uses information from the closest users, it may lose a lot of valuable ratings from a user a little further apart in the network. Massa and Avesani (2007) introduced the algorithm *MoleTrust* based on an idea similar to *TidalTrust*. But *MoleTrust* considers all users up to a maximum depth, given as an input. The maximum depth is independent of any specific user and item. Depending on the maximum depth value, *MoleTrust* may consider too many users. There are some other algorithms available which are discussed in the literature review. However, an effective and efficient trust inference algorithm which will consider all possible users as neighbors is desirable but not available. Therefore, we propose a novel trust inference algorithm which uses a modified Depth First Search (DFS). The proposed algorithm focuses on the principle of maximizing certainty in the trust value rather deriving from most positive or negative trust value.

The term "Web 2.0" was coined in January 1999 by Darcy DiNucci, a consultant on electronic information design (DiNucci 1999). The important part of Web 2.0 is the social Web, which is a fundamental shift in the way people communicate. The social Web consists of a number of online tools and platforms where people share their perspectives, opinions, thoughts and experiences. Web 2.0 applications tend to interact much more with the end user. As such, the end user is not only a user of the application but also a participant by podcasting, blogging, tagging, social bookmarking or social networking etc. Thus Web 2.0 offers a rich user information source which can be used to generate personalized recommender systems. One of the prominent features of Web 2.0 is the prevalent functionality of allowing users to annotate, describe or tag online content items in many social web sites. *Tagging*, also called social tagging, refers to the act of assigning freely-chosen descriptive keywords to online content by ordinary Web users. These tags provide rich user information. Tags can be used to profile users' topic interests or preferences because tags reflect users' perspective on how to classify items. How to use or incorporate the emerging user tagging information in Web 2.0 in personalization applications has become an interesting and important research topic.

## 1.1 Problem and Objectives

The research problem of this book is to explore novel, effective and efficient approaches to improve the quality of recommendation making by extending the neighborhood and the use of alternative information other than common rating history. In this research, we attempt to address the following two issues:

- Insufficient rating information for recommender systems
- Insufficient and unreliable trust information to generate the trust network

We set our research question broadly as "how to improve the quality of recommendation making?" Should we consider the use of alternative information other than rating history or should we extend the neighborhood or both? While making automated recommendations using the collaborative filtering technique, neighbor formation plays the major role. Whether we use previous common ratings or common interest-based trust data, the problem is finding sufficient overlap information which can be used to find the neighbors. Insufficient numbers of directly connected nodes in trust networks demands a method to infer the trust value to indirectly connect nodes and thus increase the range or boundary of the neighborhood to achieve sufficient neighbors for generating better recommendations. Therefore, to deal with these issues, an effective and efficient algorithm is required to infer trust in a network to expand the neighborhood. We have listed the following specific questions to be addressed by this book:

- How to expand the coverage or boundary of the user neighbourhood?
- How to use Web 2.0 information for user profiling?
- How to user trust data for recommendation making? and
- How to generate a trust network if explicit trust data is not available?

This research adopts the users' interest similarity as the source of trust between them for a given scope. The goal of this book is twofold. Firstly, we seek an effective way to expand the coverage or boundary of the user neighborhood to make quality recommendations. To achieve this goal we propose a novel algorithm to infer trust which is used to calculate the trust between a source node and other indirectly-connected nodes; that is, our proposed algorithm can effectively increase the number of trusted neighbors by inferring a trust value to a node which is not trusted directly but connected by a chain of trusted nodes. Based on the results obtained from the experiment conducted in the work, it has been found that the proposed techniques have achieved promising improvements in the overall recommendation making in terms of correct recommendation. Secondly, we seek to generate a trust network in the absence of explicit trust data for generating quality recommendations. The ultimate goal is to create an advanced solution which can improve the automated recommendation generation. The existing trust-based recommender works have assumed that the trust network already exists. In this work, we have assumed that interest similarity has a positive relationship with trust and this can be used to generate a trust network for recommendation. In this research, we propose a new method called *SimTrust* for developing trust networks based on user's interest similarity in the absence of explicit trust data. To identify the interest similarity, we use user's personalized tagging behavior information. Our emphasis is on what resources or the contents of those resources the user chooses to tag, rather than the textual name of the tag itself. The commonalities of the resources being tagged by the users show their common interest and this can be used to form the neighbors used in the automated recommender system. Our experimental results show that our proposed tag-similarity based method outperforms the traditional collaborative filtering approach which usually uses rating data. Thus, we argue that using the proposed approaches, the goal of more accurate, effective and efficient recommender systems can be achieved.

A highly effective and personalized recommender system may still face the challenge of data sparsity. Therefore, exploring trust inference techniques and user profiling in recommender systems by using user-generated information, especially users' tagging information is the main purpose of this book.

## 1.2 Significance and Contributions

The main focus of this book is to improve automated recommendation making with the help of trust inference and user-interest similarity based on tagging behavior. The contributions of this book can benefit research on recommender

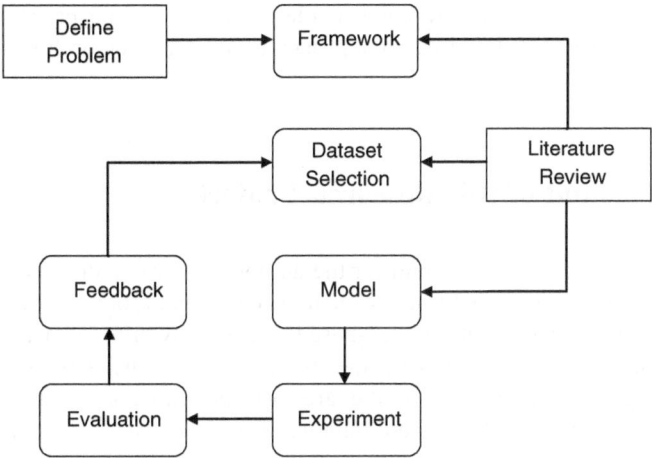

**Fig. 1.1** The proposed research method for this book

systems and online trust management. Throughout the book, we show that commonality of users' tagging behavior and inferred trust relationships offer some real benefits to the user. In order to accomplish this, we provide the following:

- New knowledge in the area of trust management through a survey on the relationship between trust and interest similarity.
- Use of a trust inference technique to increase the range/boundary of the neighbors and improve recommendation results.
- Propose a new trust inference technique which performs better than existing standard trust inference techniques.
- A new method of trust network generation in the absence of explicit trust rating using web 2.0 information such as "tag" and use it for recommendation making.

This research makes significant contributions to the field of recommender systems and online trust management since the techniques of using trust-based recommendation is highly desirable though not yet generally available.

## 1.3 Research Methodology

Various research approaches have been used in the field of recommender systems. Some of these methods include survey, case studies, prototyping and experimenting (Sarwar et al. 2000; Herlocker et al. 2004; Schafer et al. 2001). As the research is considered to focus on the development of new algorithms to improve the quality of recommendation making and assess the effectiveness of the proposed strategies, it must be supported by the results from the experiments and evaluations. Hence, the initial survey and the experimenting approach integrated with the standard information system research cycle is chosen as the proposed research method. The process of the research approach used in this research is illustrated in Fig. 1.1:

## 1.4 Background of this Research Project

This book presents the result found for the authors PhD research. The research has changed direction since its commencement. The original topic was to explore trust and reputation management in web-based social networks as part of a larger industry-linked project. However, the work has a considerable contribution to generating new knowledge in the area of recommender systems and trust management in general which is the interest of many current research student and scientists.

## 1.5 Book Outline

The book is organised as follows:

Chapter 1 introduces the background information for the research domain and presents a map of the organisation of this book.

Chapter 2 is a literature review of related trust management techniques including trust and reputation systems which leads to the research problems and questions. The relevant publication about this chapter is listed as 8, 9 and 11 in the Associated Publication list.

Chapter 3 describes the proposed trust inference techniques which are used to expand the range of the neighborhood to make trust-based recommendations. The relevant publications about this chapter are listed as 2–4, 10, 13 and 14.

Chapter 4 discusses the survey conducted on the relationship between trust and user-interest similarity which plays a vital role in the hypothesis adapted in the rest of the work. The relevant publications about this chapter are listed as 1 and 6.

Chapter 5 presents the proposed interest similarity based trust network generation algorithm which is the major contribution of this research work. The relevant publications about this chapter are listed as 5, 7 and 12.

Chapter 6 presents the experiment result with detailed discussions.

Chapter 7 the book concludes in this chapter by highlighting the contributions and outlining future work planned.

# Chapter 2
# Literature Review

**Abstract** In this literature review, we have addressed the fundamentals of trust
and reputation, in terms of an online environment. The literature review has
progressed to defining the problem domain. We have realized that it is crucial to
proactively carry out this research while online service technologies and platforms
are still being developed. It will be very difficult to start integrating soft security
solutions into online service technologies after the online service trust problems
start emerging. Online systems are currently integrated into many information
systems providing online services, where people can rate each other, or rate
products and services offered and express their feedback as opinions about products and services offered either online or through the physical world. However, the
field of online services trusts management in the recommender system is relatively
new and different, addressing the demand for automated decision support in
increasingly dynamic business to business (B2B) relationships. Some areas have
not been reviewed in depth, such as how trust can be propagated through the web-
based social network and how trust can be used to improve the automated recommendation quality. More investigation is required to develop trust management
technology for online markets, including improvement in trust inference techniques and the methods for assessing the quality of recommendation related to
trust and reliability.

**Keywords** Recommender systems · Trust · Inference · User tag · Web 2.0 ·
Social networks

In this literature review, we have addressed the fundamentals of trust and
reputation, in terms of an online environment. The literature review has progressed
to defining the problem domain. We have realized that it is crucial to proactively
carry out this research while online service technologies and platforms are still
being developed. It will be very difficult to start integrating soft security solutions
into online service technologies after the online service trust problems start
emerging. Online systems are currently integrated into many information systems
providing online services, where people can rate each other, or rate products and

T. Bhuiyan, *Trust for Intelligent Recommendation*,
SpringerBriefs in Electrical and Computer Engineering,
DOI: 10.1007/978-1-4614-6895-0_2, © The Author(s) 2013

**Table 2.1** Relationship between sections

| Section | Contents |
|---------|----------|
| 2.1 | Reviews the state of the art in trust management |
| 2.2 | Summarizes recent studies on recommender systems including the most popular collaborative filtering techniques |
| 2.3 | Introduces the concept of personalized tagging in the Web 2.0 environment |
| 2.4 | Summarizes and highlights the implications from the literature affecting this study |

services offered and express their feedback as opinions about products and services offered either online or through the physical world. However, the field of online services trusts management in the recommender system is relatively new and different, addressing the demand for automated decision support in increasingly dynamic business to business (B2B) relationships. Some areas have not been reviewed in depth, such as how trust can be propagated through the web-based social network and how trust can be used to improve the automated recommendation quality. More investigation is required to develop trust management technology for online markets, including improvement in trust inference techniques and the methods for assessing the quality of recommendation related to trust and reliability.

This book deals with the issue of trust management to improve the recommendation quality. An extensive literature review has conducted to investigate the current state of the art work on this issue in the existing literature. The organization of the literature review is presented in the Table 2.1.

In a recent work, Tavakolifard (2010) presents a categorization of the related research on trust management and similarity measurement which is closely related to our work. It has shown in the Table 2.2. We have discussed these works briefly including other state of the art research from the literature in the next 4 sections.

## 2.1 Fundamentals of Trust and Trust Network

### 2.1.1 Trust Overview

Trust has become an important topic of research in many fields including sociology, psychology, philosophy, economics, business, law and of course in IT. Far from being a new topic, trust has been the topic of hundreds of books and scholarly articles over a long period of time. Trust is a complex word with multiple dimensions. A vast body of literature on trust has grown in several area of research but it is relatively confusing and sometimes contradictory, because the term is being used with a variety of meanings (McKnight and Chervany 2002). Also a lack of coherence exists among researchers in the definition of trust. Though dozens of proposed definitions are available in the literature, we could not find a complete formal unambiguous definition of trust. In many instances, trust is used as a word

**Table 2.2** Categorization of the related work

| Paper | Web of trust | User-item rating | Rating-based similarity | Profile-based similarity |
|-------|--------------|------------------|-------------------------|--------------------------|
| Papagelis et al. 2005 | | X | X | |
| Massa and Avesani 2004 | X | | X | |
| Massa and Bhattacharjee 2004 | X | | X | |
| Massa and Avesani 2007 | X | | | |
| Avesani et al. 2004 | X | | X | |
| Avesani et al. 2005 | X | | X | |
| Weng et al. 2008 | X | | | |
| Lathia et al. 2008 | X | | | |
| Gal-Oz et al. 2008 | X | | | |
| O'Donovan and Smyth 2005 | | X | X | X |
| Ziegler and Golbeck 2007 | | X | | X |
| Ziegler and Lausen 2004 | | X | | X |
| Golbeck 2006 | | X | | X |
| Golbeck and Hendler 2006 | | X | | X |
| Golbeck 2005 | X | | | |
| Bedi and Kaur 2006 | X | | | |
| Bedi et al. 2007 | X | | | |
| Hwang and Chen 2007 | | X | | |
| Kitisin and Neuman 2006 | X | | | |
| Fu-guo and Sheng-hua 2007 | | X | | X |
| Peng and Seng-cho 2009 | X | | | X |
| Victor et al. 2008a | X | | | |
| Victor et al. 2008b | X | | | |

or concept with no real definition. Sociologists such as Deutch and Gambetta have provided some working definitions and a theoretical background to understand the basic concepts of trust and how it operates in the real world (Folkerts 2005). Hussain and Chang (2007) present an overview of the definitions of the terms of trust from the existing literature. They have shown that none of these definitions is fully capable of satisfying all of the context dependence, time dependence and the dynamic nature of trust.

The most cited definition of trust is given by Dasgupta where he defines trust as "the expectation of one person about the actions of others that affects the first person's choice, when an action must be taken before the actions of others are known" (Dasgupta 1990). This definition captures both the purpose of trust and its nature in a form that can be discussed. Deutsch (2004) states that "trusting behavior occurs when a person encounters a situation where she perceives an ambiguous path. The result of following the path can be good or bad and the occurrence of the good or bad result is contingent on the action of another person" (Hussain and Chang 2007). Another definition for trust by Gambetta is also often quoted in the literature: "trust (or, symmetrically, distrust) is a particular level of the subjective probability with which an agent assesses that another agent or group of agents will perform a particular action, both before he can monitor such action (or independently of his capacity ever to be able to monitor it) and in a context in which it affects his own action" (Gambetta 2000). But trust can be more complex than these definitions. Trust is the root of almost any personal or economic interaction. Keser states "trust as the expectation of other persons goodwill and benign intent, implying that in certain situations those persons will place the interests of others before their own" (Keser 2003). Golbeck and Hendler (2006) defines trust as "trust in a person is a commitment to an action based on belief that the future actions of that person will lead to a good outcome". This definition has a great limitation in that it considers trust as always leading to a positive outcome. But in reality, that may not be always true. Trust is such a concept that crosses disciplines and also domains. The focus of the definition differs on the basis of the goal and the scope of the projects. As mentioned earlier, in this work, we define trust as a subjective probability by which an agent can have a level of confidence in another agent in a given scope, to make a recommendation. Two general definitions of trust defined by Jøsang (Jøsang et al. 2007), which they called reliability trust (the term "evaluation trust" is more widely used by the other researchers, therefore this term is used) and decision trust respectively, are considered to define trust formally for this work. Evaluation trust can be interpreted as the reliability of something or somebody and decision trust captures a broader concept of trust.

**Evaluation Trust:** Trust is the subjective probability by which an individual, A, expects that another individual, B, performs a given action on which their welfare depends.

**Decision Trust:** Trust is the extent to which one party is willing to depend on something or somebody in a given situation with a feeling of relative security, even though negative consequences are possible.

Dimitrakos (2003) surveyed and analyzed the general properties of trust in e-services and listed the general properties of trust (and distrust) as follows:

- Trust is relevant to specific transactions only. A may trust B to drive her car but not to baby-sit.
- Trust is a measurable belief. A may trust B more than A trusts C for the same business.
- Trust is directed. A may trust B to be a profitable customer but B may distrust A to be a retailer worth buying from.
- Trust exists in time. The fact that A trusted B in the past does not in itself guarantee that A will trust B in the future. B's performance and other relevant information may lead A to re-evaluate her trust in B.
- Trust evolves in time, even within the same transaction. During a business transaction, the more A realizes she can depend on B for a service X, the more A trusts B. On the other hand, A's trust in B may decrease if B proves to be less dependable than A anticipated.
- Trust between collectives does not necessarily distribute to trust between their members. On the assumption that A trusts a group of contractors to deliver (as a group) in a collaborative project, one cannot conclude that A trusts each member of the team to deliver independently.

- Trust is reflexive, yet trust in oneself is measurable. *A* may trust her lawyer to win a case in court more than she trusts herself to do it. Self-assessment underlies the ability of an agent to delegate or offer a task to another agent in order to improve efficiency or reduce risk.
- Trust is a subjective belief. *A* may trust *B* more than *C* trusts *B* within the same trust scope.

Reputation systems are closely related to the concept of trust. The global trust in somebody or something can be referred as reputation. Wang and Vassileva (2007) identify that trust and reputation share some common characteristics such as being context specific, multi-faceted and dynamic. They argue that trust and reputation both depend on some context. Even in the same context, there is a need to develop differentiated trust in different aspects of a service. As they are dynamic in character, they say that trust and reputation can increase or decrease with further experiences of interactions or observations. Both of them also decay with time. Mui et al. (2002) differentiate the concepts of trust and reputation by defining reputation as the perception that an agent creates through past actions about its intentions and norms, and trust as a subjective expectation an agent has about another's future behavior based on the history of their encounters. The difference between trust and reputation can be illustrated by the following perfectly normal and plausible statements:

1. I trust you because of your good reputation.
2. I trust you despite your bad reputation.

Statement (1) reflects that the relying party is aware of the trustee's reputation, and bases his or her trust on that reputation. Statement (2) reflects that the relying party has some private knowledge about the trustee, e.g., through direct experience or intimate relationship, and that these factors overrule any reputation that the trustee might have. This observation reflects that trust ultimately is a personal and subjective phenomenon that is based on various factors or evidence, and that some of those carry more weight than others. Personal experience typically carries more weight than second hand recommendations or reputation, but in the absence of personal experience, trust often has to be based on reputation. Reputation can be considered as a collective measure of trustworthiness (in the sense of reliability) based on ratings from members in a community. Any individual's subjective trust in a given party can be derived from a combination of reputation and personal experience.

That an entity is trusted for a specific task does not necessarily mean that it can be trusted for everything. The scope defines the specific purpose and semantics of a given assessment of trust or reputation. A particular scope can be narrow or general. Although a particular reputation has a given scope, it can often be used as an estimate of the reputation of other scopes (Jøsang et al. 2007).

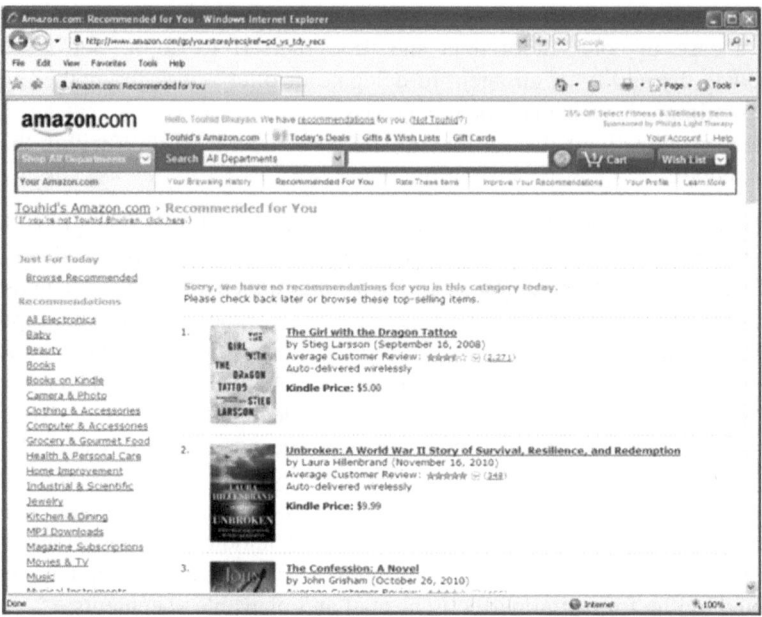

**Fig. 2.1**  The Amazon.com website

## 2.1.2  Previous Works on Trust

The issue of trust has been gaining an increasing amount of attention in a number of research communities including those focusing on online recommender systems. There are many different views of how to measure and use trust. Some researchers assign the same meaning to trust and reputation, while others do not. Though the meaning of trust is different to different people, we consider that a brief review of these models is a good starting point to research in the area of online trust management. While it is impossible to review all of the influential work related to trust management, the range of interest in the topic across nearly every academic field is impressive. These are just a few examples to give a sense of that scope.

As trust is a social phenomenon, the model of trust for an artificial world like the Internet should be based on how trust works between people in society (Abdul-Rahman and Hailes 2000). The rich and ever-growing selection of literature related to trust management systems for Internet transactions, as well as the implementations of trust and reputation systems in successful commercial applications such as eBay and Amazon (Fig. 2.1), give a strong indication that this is an important technology (Jøsang et al. 2007).

Commercial implementations seem to have settled around relatively simple principles, whereas a multitude of different systems with advanced features are being proposed by the academic community. A general observation is that the proposals from the academic community so far lack coherence and are rarely evaluated in a commercial/industrial application environment. The systems being proposed are usually designed from scratch, and only in very few cases are authors building on proposals by other authors. The period in which we are now can therefore be seen as a period of pioneers.

Stephen Marsh (1994) is one of the pioneers to introduce a computational model for trust in the computing literature. For his PhD book, Marsh investigates the notions of trust in various contexts and develops a formal description of its use with distributed, intelligent agents. His model is based on social and psychological factors. He defines trust in three categories; namely basic trust, general trust and situational trust.

These trust values are used to help an agent to decide if it is worth it or not to cooperate with another agent. Besides trust, the decision mechanism takes into account the importance of the action to be performed, the risk associated with the situation and the perceived competence of the target agent. To calculate the risk and the perceived competence, different types of trust (basic, general and situational) are used. But the model is complex, mostly theoretical and difficult to implement. He did not consider reputation in his work.

Abdul-Rahman and Hailes (2000) proposed a model for supporting trust in virtual communities, based on direct experiences and reputation. They have proposed that the trust concept can be divided into direct and recommender trust. They represent direct trust as one of four agent-specified values about another agent which they called very trustworthy, trustworthy, untrustworthy, and very untrustworthy. Recommended trust can be derived from word-of-mouth recommendations, which they consider as reputation. However, there are certain aspects of their model that are ad-hoc which limits the applicability of the model in broader scope.

Schillo et al. (2000) proposed a trust model for scenarios where the interaction result is Boolean, either good or bad, between two agents' trust relationship. They did not consider the degrees of satisfaction. They use the formula to calculate the trust that an agent $Q$ deserves to an agent $A$ is

$$T(A, Q) = \frac{e}{n} \tag{2.1}$$

where $n$ is the number of observed situations and $e$ the number of times that the target agent was honest.

Two one-on-one trust acquisition mechanisms are proposed by Esfandiari and Chandrasekharan (2001) in their trust model. The first is based on observation. They proposed the use of Bayesian networks and to perform the trust acquisition by Bayesian learning. The second trust acquisition mechanism is based on interaction. A simple way to calculate the interaction-based trust during the exploratory stage is using the formula

$$T_{\text{inter}}(A, B) = \frac{number\_of\_correct\_replies}{total\_number\_of\_replies} \qquad (2.2)$$

In the model proposed by Yu and Singh (2002), the information stored by an agent about direct interactions is a set of values that reflect the quality of these interactions. Only the most recent experiences with each concrete partner are considered for the calculations.

Yu and Singh then define their belief function to be $m$ and can use that to gauge the trustworthiness of an agent in the community. They establish both lower and upper thresholds for trust. Each agent also maintains a quality of service (QoS) metric $0 \leq s_{jk} \leq 1$ that equates to an agent $j$'s rating of the last interaction with agent $k$.

This model does not combine direct information with witness information. When direct information is available, it is considered the only source to be used to determine the trust of the target agent. Only when the direct information is not available, the model appeals to witness information.

Papagelis et al. (2005) develop a model to establish trust between users by exploiting the transitive nature of trust. However, their model simply adopts similarity as trustworthiness which still possesses the limitations of similarity based collaborative filtering. Sabater and Sierra (2005) have proposed a modular trust and reputation system oriented to complex small/mid-size e-commerce environments which they called *ReGreT*, where social relations among individuals play an important role. The system takes into account three different sources of information: direct experiences, information from third party agents and social structures. This system maintains three knowledge bases. The outcomes data base (ODB) to store previous contacts and their result; the information data base (IDB), that is used as a container for the information received from other partners and finally the sociograms data base (SDB) to store the graphs (sociograms) that define the agent's social view of the world. These data bases feed the different modules of the system. The *direct trust* module deals with direct experiences and how these experiences can contribute to the trust associated with third party agents. Together with the reputation model they are the basis on which to calculate trust. The system incorporates a credibility module that allows the agent to measure the reliability of witnesses and their information. This module is extensively used in the calculation of *witness reputation*. All these modules work together to offer a complete trust model based on direct knowledge and reputation.

Mui et al. (2002) proposed a computational model based on sociological and biological understanding. The model can be used to calculate agent's trust and reputation scores. They also identified some weaknesses in the trust and reputation study, being the lack of differentiation of trust and reputation and that the mechanism for inference between them is not explicit. Trust and reputation are taken to be the same across multiple contexts or are treated as uniform across time and the existing computational models for trust and reputation are often not grounded on understood social characteristics of these quantities. They did not examine the effects of deception in this model.

In his research, Dimitrakos (2003) presented and analysed a service-oriented trust management framework based on the integration of role-based modelling and risk assessment in order to support trust management solutions. There is also a short survey on recent definitions of trust and the authors subsequently introduce a service-oriented definition of trust, and analyse some general properties of trust in e-services, emphasising properties underpinning the inference and transferability of trust. Dimitrakos suggested that risk analysis and role-based modelling can be combined to support the formation of trust intentions and the endorsement of dependable behaviour based on trust. In conclusions, they provided evidence of emerging methods, formalisms and conceptual frameworks which, if appropriately integrated, can bridge the gap between systems modelling, trust and risk management in e-commerce. Massa and Avesani (2004, 2006) first argued that recommender system can be more effective by incorporating trust than traditional collaborative filtering. However, they used trust metric with explicit trust rating which is not easily available in real world applications. In their subsequent work (Massa and Bhattacharjee 2004) they show that the incorporation of trust metric and similarity metric can increase the coverage. The limitations of their work include the binary relationship among users and trust inference is calculated only on distance between them. Avesani et al. (2004, 2005) apply their model in ski mountaineering domain. Their model requires direct feedback from the user. They also didn't evaluate their models effectiveness.

O'Donovan and Smyth (2005) distinguished between two types of profiles in the context of a given recommendation session or rating prediction. The consumer profile and the producer profile. They described "trust" as the reliability of a partner profile to deliver accurate recommendations in the past. They described two models of trust, called profile-level trust and item-level trust. According to

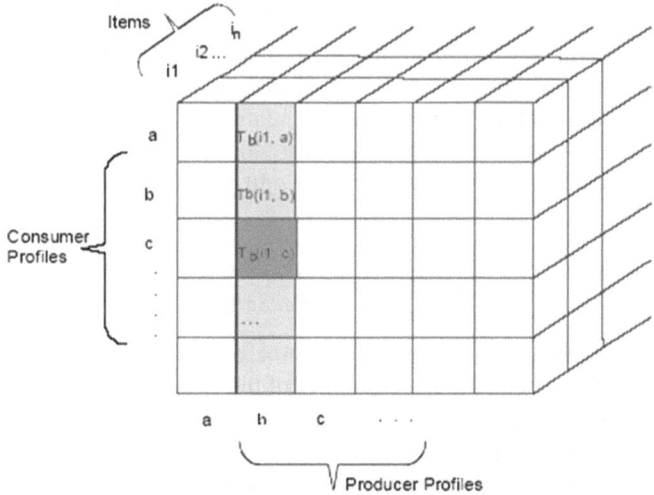

**Fig. 2.2** Calculation of trust scores from rating data

them, a ratings prediction for an item, $i$, by a producer $p$ for a consumer $c$, is correct if the predicted rating, $p(i)$, is within $\varepsilon$ of $c$'s actual rating $c(i)$; see Eq. 2.1. Of course normally when a producer is involved in the recommendation process they are participating with a number of other recommendation partners and it may not be possible to judge whether the final recommendation is correct as a result of $p$'s contribution. Accordingly, when calculating the correctness of $p$'s recommendation we separately perform the recommendation process by using $p$ as $c$'s sole recommendation partner.

For example, in Fig. 2.2, a trust score for item $i1$ is generated for producer $b$ by using the information in profile $b$ *only* to generate predictions for each consumer profile. Equation 2.8 shows how each box in Fig. 2.2 translates to a binary success/ fail score depending on whether or not the generated rating is within a distance of $\varepsilon$ from the actual rating a particular consumer has for that item. In a real-time recommender system, trust values for producers could be easily created on the fly, using a comparison between predicted rating (based only on one producer profile) and the actual rating entered by a user.

$$Correct(i,\ p,\ c) \Leftrightarrow |p(i) - c(i)| < \epsilon \qquad (2.3)$$

$$T_p(i,c) = Correct(i,\ p,\ c) \qquad (2.4)$$

From this they define two basic trust metrics based on the relative number of correct recommendations that a given producer has made. The full set of recommendations associated with a given producer, $RecSet(p)$, is given by Eq. 2.5. And the subset of these that are correct, $CorrectSet(p)$ is given by Eq. 2.6. The $i$ values represent items and the $c$ values are predicted ratings.

$$RecSet(p) = \{(c_1,\ i_1), \ldots, (c_n,\ i_n)\} \qquad (2.5)$$

$$CorrectSet(p) = \{(c_k,\ i_k) \in RecSet(p) : Correct(i_k,\ p,\ c_k)\} \qquad (2.6)$$

The profile-level trust, $Trust^P$ for a producer is the percentage of correct recommendations contributed by this producer; see Eq. 2.7. For example, if a producer has been involved in 100 recommendations, that is they have served as a recommendation partner 100 times, and for 40 of these recommendations the producer was capable of predicting a correct rating, the profile level trust score for this user is 0.4.

$$Trust^P(p) = \frac{|CorrectSet(p)|}{|RecSet(p)|} \qquad (2.7)$$

Selcuk et al. (2004) proposed a reputation-based trust management protocol for P2P networks where users rate the reliability of the parties they deal with and share this information with their peers.

Guha et al. (2004) proposed a method based on the *PageRank*$^{TM}$ algorithm for propagating both trust and distrust. They identified four different methods for propagating the net belief values, namely direct inference, co-citation, transpose and coupling. Further to these, three different methods, which they called global rounding, local rounding and majority rounding were proposed to infer whether an element in the final belief matrix, corresponds to trust or distrust.

The Advogato *MaxFlow* (maximum flow) trust metric has been proposed by Levien (2004) in order to discover which users are trusted by members of an online community and which are not. Trust is computed through one centralised community server and considered relative to a seed of users enjoying supreme trust. Local group trust metrics compute sets of agents trusted by those being part of the trust seed. In *MaxFlow* model, its input is given by an integer number $n$, which is supposed to be equal to the number of members to trust, as well as the trust seed $s$, which is a subset of the entire set of users A. The output is a characteristic function that maps each member to a Boolean value indicating his trustworthiness:

$$Trust_M : 2^A \times N_0^+ \rightarrow (A \rightarrow \{true, false\}) \tag{2.8}$$

The *AppleSeed* trust metric was proposed by Ziegler (2005). *AppleSeed* is closely based on the *PageRank*$^{TM}$ algorithm. It allows rankings of agents with respect to trust accorded. *AppleSeed* works with partial trust graph information. Nodes are accessed only when needed, i.e., when reached by energy flow. Shmatikov (Shmatikov and Talcott 2005) proposed a reputation-based trust management model which allows mutually distrusting agents to develop a basis for interaction in the absence of a central authority. The model is proposed in the context of peer-to-peer applications, online games or military situations.

Teacy (2005) proposed a probabilistic framework for assessing trust based on direct observations of a trustee's behavior and indirect observations made by a third party. Their proposed mechanism can cope with the possibility of unreliable third party information in some contexts. They also identified some key issues to consider the assessment of the reliability of reputation, like (1) assessing reputation source accuracy, (2) combining different types of evidence, which is required when there is not enough of any one type of evidence for a truster to assess a trustee; (3) exploration of trustee behavior, which involves taking a calculated risk and interacting with certain agents, to better assess their true behavior etc.

Xiong (2005) also proposed a decentralized reputation-based trust-supporting framework called *PeerTrust* for a P2P environment. Xiong focused on models and techniques for resilient reputation management against feedback aggregation, feedback oscillation and loss of feedback privacy. Xue (Xue and Fan 2008) proposed a new trust model for the Semantic Web which allows agents to decide which among different sources of information to trust and thus act rationally on the semantic web. Tian et al. (2008) proposed a trust model for P2P networks in which the trust value of a given peer was computed using its local trust information and recommendations from other nodes. A generic method for quantifying and updating the credibility of a recommender was also put forward. With their experimental results they encountered security problems in open P2P networks. Lathia et al. (2008) outlined several techniques for performing collaborative filtering on the perspective of trust management. A model for computing trust based reputation for communities of strangers is proposed by Gal-Oz et al. (2008). Bedi and Kaur (2006) proposed a model which incorporates the social recommendation process. In their latter work (Bedi et al. 2007) knowledge stored in the form of ontology is used. Hwang and Chen (2007) attempted to present their model of recommender system by incorporating trust. Victor et al. (2008a) attempted to solve the problem of cold start while making recommendation for a new user. They propose to connect a new used with underlying network for recommendation making. They also advocate the use of trust model in which trust scores are coupled (Victor et al. 2008b). Peng and Seng-cho (2009) applied their model in the domain of blog to help the readers of the blog to find desired articles easily. A recent work (Jamali et al. 2009) proposes a new algorithm called *TrustWalker* to combine trust-based and item-based recommendation. However, their proposed method is limited to centralized systems only. A trust based, effective and efficient model for recommender system is yet to develop and deserving.

## 2.2 Recommender Systems

Recommender systems intend to provide people with recommendations of items they might appreciate or be interested in. To deal with the ever-growing information overload on the Internet, recommender systems are widely used online to suggest to potential customers, items they may like or find useful. We use recommendations extensively in our daily lives. Examples of recommendations are not hard to find in our day-to-day activities. We may read movie reviews in a magazine or online hotel reviews on the Internet to decide what movies to watch or in which hotel to stay. Sometimes, we even accept recommendations from a librarian to decide which book to choose by discussing our interest and current mood. Usually, people like to seek recommendations from friends or associates when they do not have enough information to decide which books to read, movies to watch, hotels or restaurants to book etc. People love to share their preferences regarding books, movies, hotels or restaurants. Recommender systems attempt to create a technological proxy which produces the recommendation automatically

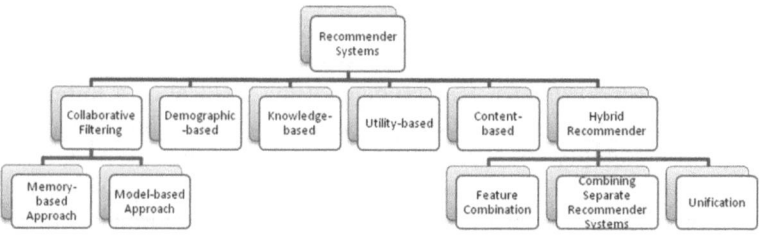

**Fig. 2.3** Taxonomy of recommender systems

based on user's previous preferences. The assumption behind many recommender systems is that a good way to produce personalized recommendations for a user is to identify people with similar interests and recommend items that may interest these like-minded people. In this section, the popularly used recommendation approaches will be outlined.

Based on the information filtering techniques employed, recommender systems can be broadly categorized in six different ways (Fig. 2.3):

- Demographic-based,
- Knowledge-based,
- Utility-based,
- Content-based,
- Collaborative filtering, and
- Hybrid recommender systems.

## 2.2.1 Demographic-Based Filtering

Demographic filtering recommender systems aim to categories the user based on personal attributes such as their education, age, occupation, and/or gender, to learn the relationship between a single item and the type of people who like it (Rich 1979; Krulwich 1997) and make recommendations based on demographic class. Grundy's system (Guttman et al. 1998) is an example of a demographic filtering recommender system which recommended books based on personal information gathered through an interactive dialogue. The users' responses were matched against a library of manually assembled user stereotypes. Hill et al. (1995) proposed a system which uses demographic groups from marketing research to suggest a range of products and services. A short survey was used to gather the data for user categorization. In Pazzani's (1999) proposed system, machine learning is used to arrive at a classifier based on demographic data. LifeStyle Finder (Krulwich 1997) is another good example of a purely demographic-filtering based recommender. LifeStyle Finder divided the population of the United States into 62 demographic clusters based on their lifestyle characteristics, purchasing history and survey responses. Hence, based on a given user's demographic information, LifeStyle Finder can deduce the user's lifestyle characteristics by finding which demographic cluster the user belongs to and make recommendations to the user. Demographic techniques form "people-to-people" correlations like collaborative filtering, but use different data. The main advantage of a demographic approach is that it may not require a history of user ratings of the type needed by collaborative and content-based techniques. However, there are some shortcomings of demographic filtering recommenders because they create user profiles by classifying users using stereotyped descriptors (Rich 1979) and recommend the same items to users with similar demographic profiles. As every user is different, these recommendations might be too general and poor in quality (Montaner et al. 2003). Purely demographic-filtering based recommenders do not provide any individual adaptation regarding interest changes (Montaner et al. 2003). However, an individual user's interests tend to shift over time, so the user profile needs to adapt to change which is generally available in collaborative filtering and content-based recommenders as both of them take users' preference data as input for recommendation making.

## 2.2.2 Knowledge-Based Filtering

Knowledge-based recommender systems use knowledge about users and products to pursue a knowledge-based approach to generating a recommendation, reasoning about what products meet the user's requirements. The *PersonaLogic* recommender system offers a dialog that effectively walks the user down a discrimination tree of product features. Others have adapted quantitative decision support tools for this task (Bhargava et al. 1999). The restaurant recommender *Entree* (Burke et al. 1997) makes its recommendations by finding restaurants in a new city similar to restaurants the user knows and likes. The system allows users to navigate by stating their preferences with respect to a given restaurant, thereby refining their search criteria. A knowledge-based recommender system does not have a ramp-up problem since its recommendations do not depend on a base of user ratings. It does not have to gather information about a particular user because its judgments are independent of individual tastes. These characteristics make knowledge-based recommenders a valuable system on their own and also complementary to other types of recommender systems. Knowledge-based approaches are distinguished from other approaches in the way that they have functional knowledge such as how a particular item meets a particular user need, and can therefore reason about the relationship between a need and a possible recommendation. The user profile can be any knowledge structure that supports this inference. In the simplest case, as in Google, it may simply be the query formulated by the user. In other cases, it may be a more detailed representation of the user's needs (Towle and Quinn 2000). Some researchers have called knowledge-based recommendation the "editor's choice" method (Schafer et al. 2001; Konstan et al. 1997). The knowledge used by a knowledge-based recommender system can also take many forms. Google uses information about the links between web pages to infer popularity and authoritative value (Brin and Page 1998). Knowledge-based recommender systems actually help users explore and thereby understand an information space. Users are an integral part of the knowledge discovery process, elaborating their information needs in the course of interacting with the system. One need only have general knowledge about the set of items and only an informal knowledge of one's needs; the system knows about the tradeoffs, category boundaries, and useful search strategies in the domain. Utility-based approaches calculate a utility value for objects to be recommended, and in principle, such calculations could be based on functional knowledge. However, existing systems do not use such inference, requiring users to do their own mapping between their needs and the features of products.

### 2.2.3 Utility-Based Filtering

Utility-based recommender systems make recommendations based on the computation of the utility of each item for the user. Utility-based recommendation techniques use features of items as background data, elicit utility functions over items from users to describe user preferences, and apply the function to determine the rank of items for a user (Burke 2002). The user profile therefore is the utility function that the system has derived for the user, and the system employs constraint satisfaction techniques to locate the best match. Utility-based recommenders do not attempt to build long-term generalisations about their users, but rather base their advice on an evaluation of the match between a user's need and the set of options available.

The e-commerce site *PersonaLogic* has different techniques for arriving at a user-specific utility function and applying it to the objects under consideration (Guttman et al. 1998). The advantage of utility-based recommendations is that they do not face problems involving new users, new items, and sparsity (Burke 2002). It can factor non-product attributes, such as vendor reliability and product availability, into the utility computation, making it possible, for example, to trade off price against delivery schedule for a user who has an immediate need. It attempts to suggest objects based on inferences about a user's needs and preferences. In some sense, all recommendation techniques could be described as doing some kind of inference. The main problem here is; how a utility function for each user should be created. The user must build a complete preference function and weigh each attribute's importance. This frequently leads to a significant burden of interaction. Therefore, determining how to make accurate recommendations with little user effort is a critical issue in designing utility-based recommender systems. Several utility-based recommender systems have been developed (Guan et al. 2002; Schmitt et al. 2002; Stolze and Stroebel 2003; Choi and Cho 2004; Lee 2004; Liu and Shih 2005; Schickel-Zuber and Faltings 2005; Manouselis and Costopoulou 2007). Most of them are based on Multi-Attribute Utility (MAU) theory. However, these methods need a considerable amount of user effort and they lack the evaluation of their preference-elicitation methods in different contexts.

### 2.2.4 Content-Based Filtering

Conventional techniques dealing with information overload use content-based filtering techniques. They analyse the similarity between items based on their contents and recommend similar items based on users' previous preferences. (Jian et al. 2005; Pazzani and Billsus 2007; Malone et al. 1987). Typically, content-based filtering techniques match items to users through classifier-based approaches or nearest-neighbor methods. In classifier-based approaches each user is associated with a classifier as a profile. The classifier takes an item as its input and then concludes whether the item is preferred by associated users based on the item contents (Pazzani and Billsus 2007). On the other hand, content-based filtering techniques based on nearest-neighbor methods store all items a user has rated in their user profile. In order to determine the user's interests in an unseen item, one or more items in the user profile with contents that are closest to the unseen item are allocated, and based on the user's preferences to these discovered neighbor items the user's preference to the unseen item can be induced (Montaner et al. 2003; Pazzani and Billsus 2007). This technique is also less affected by the cold-start problem which is one of the major weaknesses of the collaborative filtering based recommenders.

**Limitations of contents-based filtering:** It has a number of limitations over the other techniques of recommender systems which are discussed below.

**Limited applications:** The item recommended in the content-based recommender systems should be expressed as a set of features. Information retrieval techniques can effectively extract keywords as the features from a text document. However, the problem is how to extract the features from multimedia data such as audio streaming, video streaming and graphic images. In many cases, it is not possible or practical to manually assign these attributes to the items due to limitation of resources (Shardanand and Maes 1995). Another problem is that if two different items are expressed by the same set of features, how does the content-based recommender system distinguish them. This also causes inaccuracies (Shardanand and Maes 1995).

**New User Problem:** When a new user logs on to the system, and has very few preferences known, how does the content-based recommendation work? The system cannot acquire sufficient user preference. The system would not recommend an accurate item to the user. With purely content-based filtering recommenders, a user's own ratings are the only factor influencing the recommenders' performance. Hence, recommendation quality will not be very precise for users with only a few ratings (Montaner et al. 2003).

**Undue Specialisation:** People's interests vary widely. For example, if a user likes three absolutely different areas such as sport, health and finance. The problem is that the systems find it hard to extract the commonalities from these absolutely different areas. Obviously, a content-based recommendation finds it hard to cope in this situation (Ma 2008). Therefore, this technique often suffers from the over-specialization problem. They have no inherent method for generating unanticipated suggestions and therefore, tend to recommend more of what a user has already seen (Resnick and Varian 1997; Schafer et al. 2001).

**Lack of Subjective View:** Content-based filtering techniques are based on objective information about the items such as the text description of an item or the price of a product (Montaner et al. 2003), whereas a user's selection is usually based on the subjective information of the items such as the style, quality, or point-of-view of items (Goldberg et al. 1992). Moreover, many content-based filtering techniques represent item content information as word vectors and maintain no context and semantic relations among the words, therefore the resulting recommendations are usually very content centric and poor in quality (Adomavicius and Tuzhilin 2005; Burke 2002; Ferman et al. 2002; Schafer et al. 2001).

## 2.2.5  Collaborative Filtering

Collaborative filtering, one of the most popular technique for recommender systems, collects opinions from customers in the form of ratings on items, services or service providers. It is most known for its use on popular e-commerce sites such as Amazon.com or NetFlix.com (Linden et al. 2003). Essentially, a collaborative filtering based recommender automates the process of the 'word-of-mouth' paradigm: it makes recommendations to a target user by consulting the opinions or preferences of users with similar tastes to the target user (Breese et al. 1998; Schafer et al. 2001). Early recommender systems mostly utilized collaborative and content-based heuristics approaches. But over the past several years a broad range of statistical, machine learning, information retrieval and other techniques are used in recommender systems. Collaborative filtering offers technology to recommend items of potential interest to users based on information about similarities among different users' tastes. Similarity measure plays a very important role in finding like-minded users, i.e., the target user's neighborhood. The standard collaborative filtering based recommenders use a user's item ratings as the data source to generate the user's neighborhood. Because of the different taste of different people, they rate differently according to their subjective taste. If two people rate a set of items similarly, they share similar tastes. In the recommender system, this information is used to recommend items that one participant likes, to other persons in the same cluster.

There is a significant difference between collaborative filtering and reputation systems. In reputation systems, all members in the community should judge the performance of a transaction partner or the quality of a product or service consistently. Collaborative filtering takes ratings subject to taste as input, but reputation system take ratings assumed insensitive to taste as input. Collaborative filtering algorithms are generally classified into two classes: memory based and model based. The memory-based algorithm predicts the votes of the active user on a target item as a weighted average of the votes given to that item by other users. The model-based algorithm views the problem as calculating the expected value of a vote from a probabilistic perspective and uses the users' preferences to learn a model (Xiong 2005).

Since conventional collaborative filtering based recommenders usually suffer from scalability and sparsity problems, some researchers (Badrul et al. 2001; Deshpande and Karypis 2004; Linden et al. 2003) suggested a modified collaborative filtering paradigm to alleviate these problems, and this adapted approach is commonly referred to as 'item-based collaborative filtering'. The conventional collaborative filtering technique (or user-based collaborative filtering) operates based on utilizing the performance correlations among users. Unlike the user-based collaborative filtering techniques, item-based collaborative filtering techniques look into the set of items the target user has rated and compute how similar they are to the target items that are to be recommended. While content-based filtering techniques compute item similarities based on the content information of items, item-based collaborative filtering techniques determine if two items are similar by checking if they are commonly rated together with similar ratings (Deshpande and Karypis 2004).

Item-based collaborative filtering usually offers better resistance to data sparsity problem than user-based collaborative filtering. It is because in practice there are more items being rated by common users than users' rate common items (Badrul et al. 2001). Moreover because the relationship between items is relatively static, item-based collaborative filtering can pre-compute the item similarities offline whereas user-based collaborative filtering usually computes user similarities online to improve its computation efficiency. Therefore, item-based collaborative filtering is less sensitive to scalability problem (Badrul et al. 2001; Jun et al. 2006; Deshpande and Karypis 2004; Linden et al. 2003).

### 2.2.5.1 Benefits of Collaborative Filtering:

**Subjective View:** They usually incorporate subjective information about items (e.g., style, quality etc.) into their recommendations. Hence, in many cases, collaborative filtering based recommenders provide better recommendation quality than content-based recommenders, as they will be able to discriminate between a badly written and a well written article if both happen to use similar terms (Montaner et al. 2003; Goldberg et al. 1992).

**Broader Scope:** Collaborative filtering makes recommendations based on other users' preferences, whereas content-based filtering solely uses the target user's preference information. This, in turn, facilitates unanticipated recommendations because interesting items from other users can extend the target user's scope of interest beyond his or her already seen items (Sarwar et al. 2000; Montaner et al. 2003).

**Broader Applications:** Collaborative filtering based recommenders are entirely independent of representations of the items being recommended and therefore, they can recommend items of almost any type including those items from which it is hard to extract semantic attributes automatically such as video and audio files (Shardanand and Maes 1995; Terveen et al. 1997). Hence, collaborative filtering based recommenders work well for complex items such as music and movies, where variations in taste are responsible for much of the variations in preference (Burke 2002).

### 2.2.5.2 Limitations of Collaborative Filtering

**Cold-start:** One challenge commonly encountered by collaborative filtering based recommenders is the cold-start problem. Based on different situations, the cold-start problem can be characterized into two types, namely 'new-system-cold-start problem' and 'new-user-cold-start problem'. The new-system-cold-start problem refers to the circumstance where a new system has insufficient profiles of users. In this situation, collaborative filtering based recommenders have no basis upon which to recommend and hence perform poorly (Middleton et al. 2002). In the new-user-cold-start problem, recommenders are unable to make quality recommendations to new target users with no or limited rating information. This problem still happens for systems with a certain number of user profiles (Middleton et al. 2002). When a brand-new item appears in the system there is no way it can be recommended to a user until more information is obtained through another user rating it. This situation is commonly referred to as the 'early-rater problem' (Towle and Quinn 2000; Coster et al. 2002).

**Sparsity:** The coverage of user ratings can be sparse when the number of users is small relative to the number of items in the system. In other words, when there are too many items in the system, there might be many users with no or few common items shared with others. This problem is commonly referred to as the 'sparsity problem'. The sparsity problem poses a real computational challenge as collaborative filtering based recommenders may find it harder to find neighbors and harder to recommend items since too few people have given ratings (Gui-Rong et al. 2005; Montaner et al. 2003).

**Scalability:** Scalability is another major challenge for collaborative filtering based recommenders. Collaborative filtering based recommenders require data from a large number of users before being effective, as well as requiring a large amount of data from each user while limiting their recommendations to the exact items specified by those users. The numbers of users and items in e-commerce sites might increase dynamically, consequently, the recommenders will inevitably encounter severe performance and scaling issues (Sarwar et al. 2000; Gui-Rong et al. 2005).

## 2.2.6 Hybrid Recommender System

From the recommendation techniques described in previous sections, it can be observed that different techniques have their own strengths and limitations and none of them is the single best solution for all users in all situations (Wei et al. 2005). A hybrid recommender system is composed of two or more diverse recommendation techniques, and the basic rationale is to gain better performance with fewer of the drawbacks of any individual technique, as well as to incorporate various datasets to produce recommendations with higher accuracy and quality (Schafer et al. 2001). The first hybrid recommender system *Fab* was developed in mid 1990s (Balabanovi and Shoham 1997). The combination approaches can be classified into three categories according to their methods of combination: combining separate recommenders, combining features and unification (Adomavicius and Tuzhilin 2005). The active Web Museum, for instance, combines both collaborative filtering and content-based filtering to produce recommendations with appropriate aesthetic quality and content relevancy (Mira and Dong-Sub 2001). Burke (2002) has proposed taxonomy to classify hybrid recommendation approaches into seven categories, which are 'weighted', 'mixed', 'switching', 'feature combination', 'cascade', 'feature argumentation' and 'meta-level'. The central idea of hybrid recommendation is that they usually comprise of strengths from various recommendation techniques. However, it also means they might potentially include the limitations from each of those techniques. Moreover, hybrid techniques usually are more resource intensive (in terms of computation efficiency and memory usage) than stand-alone techniques, as their resource requirements are accumulated from multiple recommendation techniques. For example, a 'collaboration via content' hybrid (Pazzani 1999) might need to process both item content information and user rating data to generate recommendations, therefore will require more CPU cycles and memory than any single content-based filtering or collaborative filtering techniques.

## 2.3  Web 2.0

### 2.3.1  Social Tags

For assigning a label or organizing items, users may provide one or more keywords as a *tag* in any web site associated with Web 2.0. It is a non-hierarchical keyword or term assigned to a piece of information such as an Internet bookmark or a file name. This kind of metadata helps describe an item and allows it to be found again by browsing or searching. Tags are chosen informally and personally by the item's creator or by its viewer, depending on the system. In a tagging system, typically there is no information about the meaning or semantics of each tag. For example, the tag "tiger" might refer to the animal or the name of a football team, and this lack of semantic distinction can lead to inappropriate similarity calculation between items. People also often select different tags to describe the same item: for example, items related to the movie "Avatar" may be tagged "Movie", "Film", "Animated", "Fiction" and a variety of other terms. This flexibility allows people to classify their collections of items in the way that they find useful, but the personalized variety of terms can make it difficult for people to find comprehensive information about a subject or item. Another common challenge in tagging systems is that people use both singular and plural words as tags. A user could tag an object with "dinosaur" or with "dinosaurs", which can make finding similar objects more difficult for both that user and other users in the system. However, user tags can be regarded as expressions of a user's personal opinion or considered as an implicit rating on the tagged item. Thus tagging information could be useful while making recommendations (Halpin et al. 2007).

Although there are a good number of works are available in the field of recommender systems using collaborative filtering, very few researchers consider using tag information to make recommendation (Tso-Sutter et al. 2008; Liang et al. 2009a). Tso-Sutter et al. used tag information as a supplementary source to extend the rating data, not to replace the explicit rating information. Liang et al. (2009a) proposed to integrate social tags with item taxonomy to make personalised user recommendations. Other recent works include integrating tag information with content-based recommender systems (Gemmis et al. 2008), extending the user-item matrix to user-item-tag matrix to collaborative filtering item recommendations (Liang et al. 2009b), combining users' explicit ratings with the predicted users' preferences for items based on their inferred preferences for tags (Sen et al. 2009) etc. However, using tags for recommender systems is still in demand.

The Web 2.0 applications attract users to provide, create, and share more information online. The current popularly available online information that is contributed by users can be classified into textural content information, multimedia content information, and friends/network information. The popular textural user contributed content information includes tags, blogs, reviews, micro-blogs, comments, posts, documents, Wikipedia articles and others. Besides textural contents, user contributed video clips, audio clips, and photos are also popularly available on the web now. In addition, users' network relationship information such as "trust", "tweets", "friends", "followers", etc. becomes more and more popular. Compared with web logs, these kinds of new user information have the advantages of being lightweight, small sized and explicitly and proactively provided by users. They become important information sources to online users.

Tag information is kind of typical Web 2.0 information. Recently, social tags become an important research focus. Implying users' explicit topic interests, social tags can be used to improve searching (Bao et al. 2007; Begelman et al. 2006; Bindelli et al. 2008), clustering (Begelman et al. 2006; Shepitsen et al. 2008), and recommendation making (Niwa et al. 2006; Shepitsen et al. 2009; Jaschke et al. 2007; Zhang et al. 2010).

The research of tags is mainly focusing on how to build better collaborative tagging systems (Golder et al. 2006), item navigation and organisation (Bindelli et al. 2008), semantic cognitive modelling in social tagging systems (Fu et al. 2009), personalising searches using tag information (Bao et al. 2007) and recommending tags (Marinho and Schmidt-Thieme 2007) and items (Tso-Sutter et al. 2008) to users etc. In real life, collaborative tagging systems have not only been used in social sharing websites such as De.licio.us, and e-commerce websites such as Amazon.com, but also in other traditional websites or organizations such as libraries and enterprises (Millen et al. 2006, 2007).

In those early works, only the original tag set of a user was used to profile the topic interests of that user (Bogers and Bosch 2008). The binary weighting approach (Bogers and Bosch 2008) and the *tf–idf* weighting approach (Salton and Buckley 1988) that borrowed from text mining are commonly used to assign different weights to tags. Tso-Sutter et al. (2008) proposed a tag expansion method to profile a user with tags and items. They proposed to convert the three-dimensional user-tag-item relationship into an expanded user-item implicit rating matrix. The tags of each user were regarded as special items to expand each user's item set. While the tags of each item were regarded as special users to expand each item's user set. This approach profiles users with their own tags and items.

Because of the noise of tags, some user profiling approaches proposed to use the related or expanded tags to profile each user or describe each item. In the work of Niwa et al. (2006) and Shepitsen et al. (2009), those tags of the same cluster were used to expand the tag-based user profiles. Yeung et al. (2008) proposed to find a cluster of tags to profile the topic interests of each user, called *personomy*. All the popular tags of the collected items of a user were used to profile the topic interests of that user. Moreover, association rule mining techniques were used to find the associated tags to profile users (Heymann et al. 2008). Wetzker et al. (2010) proposed a probability model to translate a user's tag vocabulary into another user's tag vocabulary. Thus, these approaches not only just profiled users with their own tags but also with a set of related or expanded tags.

Moreover, targeting the tag quality problem, some latent semantic models were proposed to process tags. The tag-based latent topics were used to profile users. Wetzker et al. (2009) proposed a hybrid probabilistic latent semantic model based on the binary user-item and tag-item matrixes. Siersdorfer and Sizov (2009) proposed a latent Dirichlet Analysis model to find the latent topics of tags. Thus, instead of using tags directly, these approaches used the latent topics of tags to profile users. Besides being used as a stand-alone information source, tags are also popularly used to combine with other information sources to profile users. In the next two subsections, firstly the user profiling approaches based on other popular Web 2.0 information sources will be briefly reviewed. Then, the hybrid user profiling approaches based on tags and other information sources will be reviewed.

Tags are also popularly used to combine with other information sources to profile users. Sen et al. (2009) proposed to combine tags, explicit ratings, implicit ratings, click streams and search logs to profile users and make personalized item recommendations. The work in (Heymann et al. 2008; Gemmis 2008) proposed to combine tags with the textural content information of tagged items to find users' topic interests. The work proposed approaches to combine tags with blogs (Hayes et al. 2007; Qu et al. 2008) to find users' opinions and topic interests. Recently, some approaches combined tags and images (Xu et al. 2010) and videos (Chen et al. 2010) to mine the semantic meaning of the multimedia items. The research of user profiling approaches that combine tags and micro-blogs such as tweets (e.g., hash tag) is one of the new research focuses (Huang et al. 2010; Efron 2010).

In conclusion, user profiling is an ongoing research area. The new user information provides new opportunities to profile users. At the same time, the noise and the new features of these kinds of user information bring challenges to the current user profiling approaches. How to reduce the noise, make use of the unique features and the rich personal information of these kinds of user contributed information is vital to accurately profile users.

## 2.3.2  Web 2.0 and Social Networks

A Web 2.0 site allows users to interact and collaborate with each other in a social media dialogue as creators of user-generated content in a virtual community, in contrast to websites where users are limited to the passive viewing of content that was created for them. Examples of Web 2.0 include social networking sites, blogs, wikis, video sharing sites, hosted services and web applications. However, social networking has been around for some time. Facebook and MySpace have become iconic and other sites such as LinkedIn, hi5, Bebo, CyWorld and Orkut are becoming important as well. At the end of 2007, Microsoft paid $240 million for a 1.6 % stake in Facebook, sparking a fierce debate about the theoretical valuation of Facebook. While few would go along with the $15 billion price tag, nobody would deny the huge potential of Facebook. The relevance of social networking for advertisers is very high considering they want to invest their money where their potential customers are located on social networking sites.

The success of social networking should not come as a surprise. Social interaction is deeply rooted in human nature and is one of the most fundamental needs. Wireless and Internet technology act as enablers and facilitators for enhanced social interaction with a global reach. While social networking has been and still is dominated by teenagers and young adults, it is quickly spreading to all age groups and beyond the confines of consumer entertainment. Corporations are discovering the power of networking sites to enhance their brands, communities, and overall interaction with their customers by seamlessly linking corporate Web sites to public sites such as Facebook. And something even bigger is about to take place.

There has been dramatic growth in the number and size of Web-based social networks. The number of sites almost doubled over the two-year period from December 2004 to December 2006, growing from 125 to 223. Over the same period, the total number of members among all sites grew fourfold from 115 million to 490 million (Golbeck 2007). This growth has continued over the last five years to an even greater extent. A list of the current most popular social networks can be found at http://en.wikipedia.org/wiki/List_of_social_networking_websites and also at http://trust.mindswap.org.

The recent emergence of location-based mobile social networking services offered by providers such as Rummble, GyPSii, Whrrl and Loopt is revolutionising social networking, allowing users to share real-life experiences via geo-tagged user-generated multimedia content, see where their friends are and meet up with them. This new technology-enabled social geo-lifestyle will drive the uptake of location-based services and provide opportunities for location-based advertising in the future.

A location-based social network is a system in the form of a robust web service, used to build an open infrastructure to introduce and connect individuals based on the intersection of physical location and other properties they might have in common. It is different from the wide range of existing social networking and instant messaging applications in terms of its basic activities. Location-based social networking is really all about the physical presence of a member, and feeds that information nicely into social networking applications such as Facebook, MySpace etc. Location-based social networking is a natural extension to the mobile devices of the Web-based versions of these major social networking sites. It is the mobile's ability to provide real-time location, and in turn actual presence ("I'm shopping in Brisbane") that will provide a real boost to the mobile versions of these sites. Instead of just being a cut-down version of the main site, the mobile version of a location-based social network adds real time value with presence from location services. The location-based social network can be a good application area of trust-based recommender system. Members of the network may receive recommendations from their trusted friends while they are on visiting a new place.

## 2.4 Chapter Summary

This chapter analyzed a wide range of literature relevant to trust, trust networks, recommender systems, and use of personalized tags in Web 2.0 environments. From the literature, it is understandable that trust is a way of doing social personalization. There are a good number of models proposed to calculate trust between users in a trust network while the trusts between users' data are present. It leads to research questions such as: if there is a social network with no trust values, how do we compute trust; how do we compute indirect trust, while finding the neighbors in trust networks; and how many hops should we traverse or how many of the raters should be considered, etc. If we use too few, the prediction is not based on a significant number of ratings. On the other hand, if we use too many, these raters may only be weakly trusted. In a large trust network, efficiency of exploring the trust network is another issue. It can be noted that all the relevant existing works have assumed that a trust network already exists or that the trust data is readily available, which is not very common in real world. We have considered such a situation in our proposed model which is described in detail in Chap. 5.

# Chapter 3
# Trust Inferences Using Subjective Logic

**Abstract** This chapter introduces Subjective Logic with its notations and definitions followed by the transitivity property of trust and its use in trust inference algorithms. The proposed trust inference algorithm using subjective logic is also described in this chapter. The objective of this part of work is to develop an algorithm for inferring a trust value from one node to another where no direct connection exists between the nodes in a trust network. Trust Inference can be used to recommend to one node how much to trust another node in the network. In other word, what trust rating one person wants to give another unknown person if there is an indirect link present? Due to the limited number of ratings that exist in recommender systems, underlying networks that are generated based on user rating similarity are very sparse. There are cases in which insufficient or lack of information is detrimental for the recommendation algorithm.

**Keywords** Recommender systems · Trust · Inference · User tag · Web 2.0 · Social networks

This chapter introduces Subjective Logic with its notations and definitions followed by the transitivity property of trust and its use in trust inference algorithms. The proposed trust inference algorithm using subjective logic is also described in this chapter. The objective of this part of work is to develop an algorithm for inferring a trust value from one node to another where no direct connection exists between the nodes in a trust network. Trust Inference can be used to recommend to one node how much to trust another node in the network. In other word, what trust rating one person wants to give another unknown person if there is an indirect link present? Due to the limited number of ratings that exist in recommender systems, underlying networks that are generated based on user rating similarity are very sparse. There are cases in which insufficient or lack of information is detrimental for the recommendation algorithm.

The trust-based recommendation is very similar to prediction generated by standard recommender systems. Any trust inference algorithm should follow the basic properties of trust such as transitivity, combinability and personalization. Transitivity allows information to be passed along paths back from destination to

T. Bhuiyan, *Trust for Intelligent Recommendation*,
SpringerBriefs in Electrical and Computer Engineering,
DOI: 10.1007/978-1-4614-6895-0_3, © The Author(s) 2013

source in the network and combinability refers to ability of the source to aggregate information from multiple sources. Just like trust itself, trust algorithms also have two types; global and local algorithms. The first one is used to compute a universal trust value for each node in the network. No matter which node for which the trust value is made, the value will be the same for any node within the network. But the second type of algorithm is used to calculate a personalized trust value for each node. In this case, the result value will be different depending on who is asking or for whom the trust is generated. As trust is personal in nature and it may vary between two persons, personalized trust computation through local trust algorithms should improve the accuracy of the result. A global trust represents the opinion of many entities on an object but it does not represent the suggestion of an individual. Our proposed trust inference algorithm is based on local trust and totally on the perspective of individual users. It considers friends and other members of their friends' networks whom the user trusts about their opinions on a topic. While doing this, the people for whom the user has more trust are given more emphasis than the users whom they trust less. The following sections are organized as follows. Firstly a concise notation is defined with which trust transitivity and parallel combination of trust paths can be expressed. Secondly it defines a method for simplifying complex trust networks so that they can be expressed in this concise form. Then it also allows trust measures to be expressed as beliefs, so that derived trust can be automatically and securely computed with subjective logic. In addition to the belief, we have also considered disbelief, confidence and uncertainty about the opinion in our proposed algorithm.

## 3.1 Introduction to Subjective Logic

Jøsang (2001) introduces the subjective logic which is an attempt to overcome the limits of the classical logic by taking in consideration the uncertainty, the ignorance and the subjective characteristics of the beliefs. Subjective logic is a belief calculus specifically developed for modeling trust relationships. It is a type of probabilistic logic that allows probability values to be expressed with degrees of uncertainty. Probabilistic logic combines the strengths of logic and probability calculus. It has binary logic's capacity to express structured argument models and it has also the power of probabilities to express degrees of truth of those arguments. Subjective logic makes it possible to express uncertainty about the probability values themselves by reasoning with argument models in the presence of uncertain or partially incomplete evidence. Subjective logic has a sound mathematical basis and it defines a rich set of operators for combining subjective opinions in various ways (Jøsang 2010). Some operators represent generalizations of binary logic and probability calculus, whereas others are unique to belief calculus because they depend on belief ownership. With belief ownership it is possible to explicitly express that different agents have different opinions about the same issue. The advantage of subjective logic over probability calculus and binary logic is its ability to explicitly express and take advantage of ignorance and belief ownership. Subjective logic can be applied to all situations where probability calculus can be applied, and also to many situations where probability calculus fails precisely because it cannot capture degrees of ignorance. It defines a number of operators. Some operators represent generalizations of binary logic and probability calculus operators, whereas others are unique to belief theory because they depend on belief ownership. Here, we will focus on the *discounting* and *cumulative fusion* operators. The *discounting* operator can be used to derive trust from a trust path consisting of a chain of trust edges, and the *cumulative fusion* operator can be used to combine trust from parallel trust paths. The detailed techniques of serial and parallel trust calculation are described in Sect. 3.3.

### 3.1.1 Subjective Opinion Representation

Subjective opinions express subjective beliefs about the truth of propositions with degrees of uncertainty and can indicate subjective belief ownership whenever required. Opinions can be binomial or multinomial. A multinomial opinion is denoted as $\omega_X^A$ where $A$ is the belief owner or the subject and $X$ is the target state space, to which the opinion applies. Opinions over binary state spaces are called binomial opinions. In the case of binomial opinions, the notation is $\omega_x^A$ where $x$ is a single proposition that is assumed to belong to a state space denoted as $X$, but the space is usually not included in the notation for binomial opinions. The belief owner (subject) and the propositions (object) are attributes of opinion. A general $n$-ary space $X$ can be considered binary when seen as a binary partition consisting of one of its proper subset $x$ and the complement $\bar{x}$ (Jøsang 2009, 2010).

**Definition 3.1** (*Binomial Opinion*) Let $X = \{x, \bar{x}\}$ be either a binary space or a binary partitioning of an $n$-ary space. A binomial opinion about the truth of state $x$ is the ordered quadruple $\omega_x = (b, d, u, a)$ where:

$b$   denotes belief which *is the belief mass in support of x being true,*
$d$   denotes disbelief which *is the belief mass in support of x being false,*
$u$   denotes uncertainty which *is the amount of uncommitted belief mass,*
$a$   denotes base rate which *is the* a priori *probability in the absence of committed belief mass.*

These components satisfy $b + d + u = 1$ and $b, d, u, a \in [0, 1]$. The characteristics of various binomial opinion classes are listed below.

A binomial opinion:

$b = 1$ is equivalent to binary logic TRUE,
$d = 1$ is equivalent to binary logic FALSE,
$b + d = 1$ is equivalent to a traditional probability,
$b + d < 1$ expresses degrees of uncertainty, and
$b + d = 0$ expresses total uncertainty.

**Fig. 3.1** Opinion triangle
with example opinion

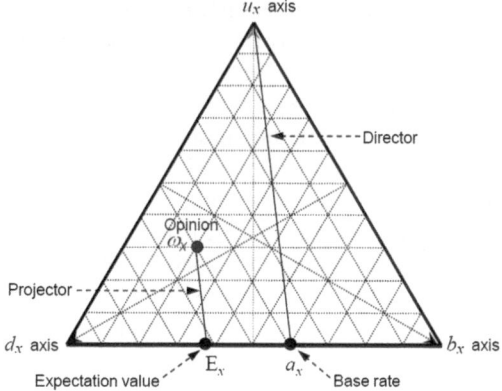

The probability expectation value of a binomial opinion is defined as $E_x = b + au$. Binomial opinions can be represented on an equilateral triangle as shown in Fig. 3.1. A point inside the triangle represents a $(b, d, u)$ triple. The belief, disbelief, and uncertainty-axes run from one edge to the opposite vertex indicated by the $b_x$ axis, $d_x$ axis and $u_x$ axis labels.

For example, a strong positive opinion is represented by a point towards the bottom right belief vertex. The base rate 1, is shown as a point on the base line, and the probability expectation, $E_x$, is formed by projecting the opinion point onto the base, parallel to the base rate director line. The opinion $\omega_x = (0.2, 0.5, 0.3, 0.6)$ with expectation value $E_x = 3.8$ is shown as an example.

When the opinion values are constrained to one of the three vertices, i.e. with $b = 1, d = 1$ or $u = 1$, subjective logic becomes a form of three-valued logic that is compatible with Kleene logic (Fitting 1994). However, the three-valued arguments of Kleene logic do not contain base rates, so that the expectation value of binomial opinions cannot be derived from Kleene logic arguments.

### 3.1.2 Operators of Subjective Logic

Table 3.1 provides a brief overview of the basic subjective logic operators. Additional operators exist for modelling special situations, such as when fusing opinions of multiple observers. Most of the operators correspond to well-known operators from binary logic and probability calculus, whereas others are specific to subjective logic.

**Addition Operator:** The addition of opinions in subjective logic is a binary operator that takes opinions about two mutually exclusive alternatives (i.e. two disjoint subsets of the same space) as arguments, and outputs an opinion about the union of the subsets. The operator for *addition* is defined.

**Table 3.1** Correspondence between subjective logic and binary logic/set operators

| Subjective logic operator | Symbol | Binary logic/set operator | Symbol | Subjective logic notation |
|---|---|---|---|---|
| Addition | + | XOR | ∪ | $\omega_{x \cup y} = \omega_x + \omega_y$ |
| Subtraction | − | Difference | \ | $\omega_{x \setminus y} = \omega_x - \omega_y$ |
| Multiplication | . | AND | ∧ | $\omega_{x \wedge y} = \omega_x \cdot \omega_y$ |
| Division | / | UN-AND | $\overline{\wedge}$ | $\omega_{x \overline{\wedge} y} = \omega_x / \omega_y$ |

**Definition 3.2** (*Addition*) Let $x$ and $y$ be two disjoint subsets of the same space $X$, i.e. $x \cap y = \emptyset$. The opinion about $x \cup y$ as a function of the opinions about $x$ and $y$ is defined as:

$$\omega_{x \cup y} : \begin{cases} b_{x \cup y} = b_x + b_y \\ d_{x \cup y} = \dfrac{a_x(d_x - b_y) + a_y(d_y - b_x)}{a_x + a_y} \\ u_{x \cup y} = \dfrac{a_x u_x + a_y u_y}{a_x + a_y} \\ a_{x \cup y} = a_x + a_y \end{cases} \quad (3.1)$$

By using the symbol "+" to denote the addition operator for opinions, addition can be denoted as $\omega_{x \cup y} = \omega_x + \omega_y$.

**Subtraction Operator:** The inverse operation to *addition* is *subtraction*. Since addition of opinions yields the opinion about $x \cup y$ from the opinions about disjoint subsets of the state space, then the difference between the opinions about $x$ and $y$ (i.e. the opinion about $x \backslash y$) can only be defined if $y \subseteq x$ where $x$ and $y$ are being treated as subsets of the state space $X$, i.e. the system must be in the state $x$ whenever it is in the state $y$. The operator for *subtraction* is defined.

**Definition 3.3** (*Subtraction*) Let $x$ and $y$ be subsets of the same state space $X$ so that $x$ and $y$, i.e. $x \cap y = y$. The opinion about $x \backslash y$ as a function of the opinions about $x$ and $y$ is defined as:

The opinion about $x \backslash y$ is given by

$$\omega_{x \backslash y} : \begin{cases} b_{x \backslash y} = b_x - b_y \\ d_{x \backslash y} = \dfrac{a_x(d_x + b_y) - a_y(1 + b_y - b_x - u_y)}{a_x - a_y} \\ u_{x \backslash y} = \dfrac{a_x u_x - a_y u_y}{a_x - a_y} \\ a_{x \backslash y} = a_x - a_y \end{cases} \quad (3.2)$$

Since $u_{x \backslash y}$ should be non-negative, then this requires that $a_y u_y \le a_x u_x$, and since $d_{x \backslash y}$ should be non-negative, then this requires that $a_x (d_x + b_y) \ge a_y (1 + b_y - b_x - u_y)$. By using the symbol "−" to denote the subtraction operator for opinions, subtraction can be denoted as $\omega_{x \backslash y} = \omega_x - \omega_y$.

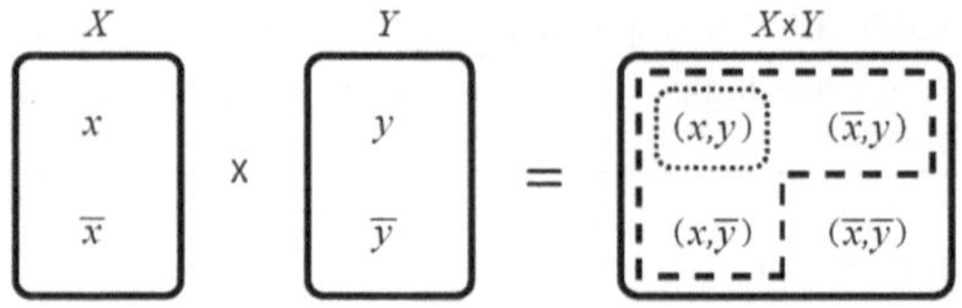

**Fig. 3.2** Cartesian product
of two binary state spaces of
discernment

**Binomial Multiplication Operator:** Binomial *multiplication* in subjective logic takes binomial opinions about two elements from distinct binary state spaces of discernment as input arguments and produces a binomial opinion as the result. The product result opinions relate to subsets of the Cartesian product of the two binary state spaces of discernment.

The Cartesian product of the two binary state spaces of discernment $X = \{x, \bar{x}\}$ and $Y = \{y, \bar{y}\}$ produces the quaternary set $X \times Y = \{(x, y), (x, \bar{y}), (\bar{x}, y), (\bar{x}, \bar{y})\}$ which is illustrated in Fig. 3.2.

As explained below, binomial multiplication in subjective logic represents approximations of the analytically correct product of Beta probability density functions (Jøsang and McAnally 2004). In this regard, normal multiplication produces the best approximations.

Let $\omega_x$ and $\omega_y$ be opinions about $x$ and $y$ respectively held by the same observer. Then the product opinion $\omega_{x \wedge y}$ is the observer's opinion about the conjunction $x \wedge y = \{(x, y)\}$ that is represented by the area inside the dotted line in Fig. 3.2.

**Definition 3.4** (*Normal Binomial Multiplication*) Let $X = \{x, \bar{x}\}$ and $Y = \{y, \bar{y}\}$ be two separate state spaces, and let $\omega_x = (b_x, d_x, u_x, a_x)$ and $\omega_y = (b_y, d_y, u_y, a_y)$ be independent binomial opinions on $x$ and $y$ respectively. Given opinions about independent propositions, $x$ and $y$, the binomial opinion $\omega_{x \wedge y}$ on the conjunction $(x \wedge y)$ is given by

$$\omega_{x \wedge y} : \begin{cases} b_{x \wedge y} = b_x b_y + \dfrac{(1 - a_x)a_y b_x u_y + a_x(1 - a_y)u_x b_y}{1 - a_x a_y} \\[2mm] d_{x \wedge y} = d_x + d_y - d_x d_y \\[2mm] u_{x \wedge y} = u_x u_y + \dfrac{(1 - a_y)b_x u_y + (1 - a_x)u_x b_y}{1 - a_x a_y} \\[2mm] a_{x \wedge y} = a_x a_y \end{cases} \tag{3.3}$$

By using the symbol "·" to denote this operator, multiplications of opinions can be written as $\omega_{x \wedge y} = \omega_x \cdot \omega_y$.

**Binomial Division Operator:** The inverse operation to binomial *multiplication* is binomial *division*. The quotient of opinions about propositions $x$ and $y$ represents the opinion about a proposition $z$ which is independent of $y$ such that $\omega_x = \omega_{y \wedge z}$. This requires that:

$$\begin{cases} a_x < a_y \\ d_x \geq d_y \\ b_x \geq \dfrac{a_x(1 - a_y)(1 - d_x)b_y}{(1 - a_x)a_y(1 - d_y)} \\ u_x \geq \dfrac{(1 - a_y)(1 - d_x)u_y}{(1 - a_x)(1 - d_y)} \end{cases} \tag{3.4}$$

**Definition 3.5** (*Normal Binomial Division*) Let $X = \{x, \bar{x}\}$ and $Y = \{y, \bar{y}\}$ be state spaces, and let $\omega_x = (b_x, d_x, u_x, a_x)$ and $\omega_y = (b_y, d_y, u_y, a_y)$ be binomial opinions on $x$ and $y$ satisfying Eq. (3.4) above. The division of $\omega_x$ by $\omega_y$ produces the quotient opinion $\omega_{x \wedge y} = (b_{x \wedge y}, d_{x \wedge y}, u_{x \wedge y}, a_{x \wedge y})$ defined by

$$\omega_{x \bar{\wedge} y} : \begin{cases} b_{x \bar{\wedge} y} = \dfrac{a_y(b_x + a_x u_x)}{(a_y - a_x)(b_y + a_y u_y)} - \dfrac{a_x(1 - d_x)}{(a_y - a_x)(1 - d_y)} \\ d_{x \bar{\wedge} y} = \dfrac{d_x - d_y}{1 - d_y} \\ u_{x \bar{\wedge} y} = \dfrac{a_y(1 - d_x)}{(a_y - a_x)(1 - d_y)} - \dfrac{a_y(b_x + a_x u_x)}{(a_y - a_x)(b_y + a_y u_y)} \\ a_{x \bar{\wedge} y} = \dfrac{a_x}{a_y} \end{cases} \tag{3.5}$$

By using the symbol "/" to denote this operator, division of opinions can be written as $\omega_{x \wedge y} = \omega_x / \omega_y$.

Subjective logic is a generalization of binary logic and probability calculus. This means that when a corresponding operator exists in binary logic, and the input parameters are equivalent to binary logic TRUE or FALSE, then the result opinion is equivalent to the result that the corresponding binary logic expression would have produced. We have considered the case of binary logic AND which corresponds to *multiplication* of opinions. Similarly, when a corresponding operator exists in probability calculus, then the probability expectation value of the result opinion is equal to the result that the corresponding probability calculus expression would have produced with input arguments equal to the probability expectation values of the input opinions.

## 3.2 Transitivity and Trust Inferences

Trust transitivity means, for example, that if $A$ trusts $B$ who trusts $C$, then $A$ will also trust $C$. This assumes that $A$ is actually aware that $B$ trusts $C$. This could be achieved through a suggestion from $B$ to $A$ as illustrated in Fig. 3.3, where the

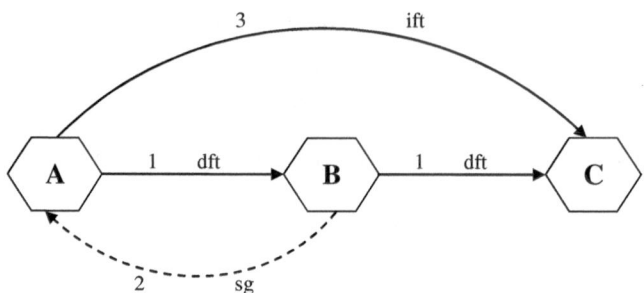

ift – Indirect Functional Trust
dft – Direct Functional Trust
sg – Suggestion

**Fig. 3.3**  Trust transitivity

indexes on each arrow indicate the sequence in which the trust relationships/
suggestion is formed.

The "dft" denotes direct functional trust which means trust generated based on
firsthand experience, "ift" denotes indirect functional trust which means trust
derived through a chain of friends' suggestions, "drt" denotes direct referral trust
and "sg" denotes suggestions from a friend.

Trust is not always transitive in real life (Christianson 2003). For example the
fact that $A$ trusts $B$ to look after her child, and $B$ trusts $D$ to fix his car, does not
imply that $A$ trusts $D$ to look after her child, or to fix her car. However, under
certain semantic constraints (Jøsang and Pope 2005), trust can be transitive, and a
trust system can be used to derive trust. In this example, trust transitivity collapses
because the scopes of $A$'s and $B$'s trust are different. Trust scope is defined as the
specific type(s) of trust assumed in a given trust relationship.

It is important to differentiate between trust in the ability to suggest a good car
mechanic, which represents *referral trust*, and trust in actually being a good car
mechanic, which represents *functional trust*. Here, functional trust refers to the
trust based on first hand personal experience and referral trust refers to the trust
derived from someone's suggestions or reference. The scope of the trust is nev-
ertheless the same, e.g. to be a good car mechanic. Assuming that, on several
occasions, $B$ has proved to $A$ that he is knowledgeable in matters relating to car
maintenance, $A$'s referral trust in $B$ for the purpose of suggesting a good car
mechanic can be considered to be *direct*. Assuming that $D$ on several occasions
has proved to $B$ that he is a good mechanic, $B$'s functional trust in $D$ can also be
considered to be direct. Thanks to $B$'s advice, $A$ also trusts $D$ to actually be a good
mechanic. However, this functional trust must be considered to be *indirect*,
because $A$ has not directly observed or experienced $D$'s skills in car mechanics. Let
us slightly extend the example, wherein $B$ does not actually know any car
mechanics himself, but he knows $C$, whom he believes knows a good car
mechanic. As it happens, $C$ is happy to suggest the car mechanic named $D$. As a

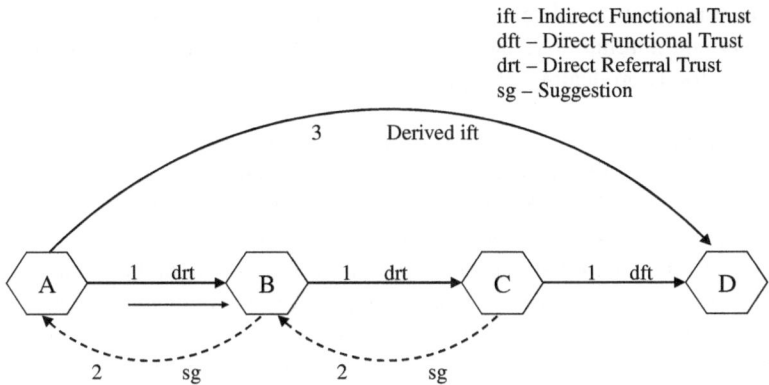

**Fig. 3.4** Serial trust path

result of transitivity, $A$ is able to derive trust in $D$, as illustrated in Fig. 3.4, where the indexes indicate the order in which the trust relationships and suggestions are formed. The "referral" variant of a trust scope can be considered to be recursive, so that any transitive trust chain, with arbitrary length, can be expressed using only one trust scope with two variants. This principle which can be expressed as the derivation of functional trust through referral trust, requires that the last trust edge represents functional trust and all previous trust edges represent referral trust. It could be argued that negative trust in a transitive chain can have the paradoxical effect of strengthening the derived trust.

Take for an example the case of Fig. 3.3, but in this case $A$ distrusts $B$, and $B$ distrusts $D$. In this situation, $A$ might actually derive positive trust in $D$, since she does not believe $B$ when he says: "*D is bad mechanic, do not use him*". So the fact that $B$ recommends distrust in $D$ might count as a pro-$D$ argument from $A$'s perspective. The question boils down to "*is the enemy of my enemy my friend?*"

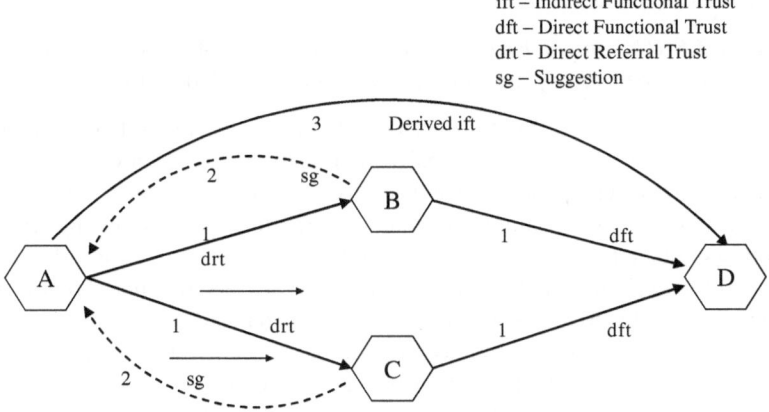

**Fig. 3.5** Parallel trust paths

However this question relates to how multiple types of untrustworthiness, such as dishonesty and unreliability, should be interpreted in a trust network. In this book, we did not focus on dealing with distrust. A low value of trust represents distrust here.

It is common to collect advice from several sources in order to be better informed when making decisions. This can be modeled as *parallel trust combination* illustrated in Fig. 3.5, where again the indexes indicate the order in which the trust relationships and recommendations are formed.

Let us assume again that $A$ needs to get her car serviced, and that she asks $B$ to suggest a good car mechanic. When $B$ suggests $D$, $A$ would like to get a second opinion, so she asks $C$ whether she has heard about $D$. Intuitively, if both $B$ and $C$ suggest $D$ as a good car mechanic, $A$'s trust in $D$ will be stronger than if she had only asked $B$.

Parallel combinations of positive trust thus have the effect of strengthening the derived trust. In the case where $A$ receives conflicting suggested trust, e.g. trust and distrust at the same time, she needs some method for combining these conflicting suggestions in order to derive her trust in $D$. Our method, which is described in the next sections, is based on subjective logic which easily can handle such cases. Subjective logic is suitable for analysing trust networks because trust relationships can be expressed as subjective opinions with degrees of uncertainty. For details about Subjective Logic, please refer to Jøsang (2010).

## 3.3 Trust Network Analysis with Subjective Logic

In this section we describe our model for trust derivation with Subjective Logic. Subjective logic represents a practical belief calculus which can be used for calculative analysis of trust networks. Our proposed approach requires trust relationships to be expressed as beliefs, and trust networks to be expressed as Directed Series Parallel Graphs (DSPG) in the form of canonical expressions.

**Definition 3.6** (*Canonical Expression*) An expression of a trust graph in structured notation where every arc only appears once is called canonical.

In this section, we describe how trust can be derived with the belief calculus of subjective logic. Trust networks consist of transitive trust relationships between nodes which represent people, organisations or software agents, etc. connected through a medium for communication and interaction. By formalising trust relationships as subjective trust measures, trust between parties within a domain can be derived by analysing the trust paths linking the parties together. Trust Network Analysis using Subjective Logic (TNA-SL) was introduced by Jøsang et al. (2006). We have analyzed the algorithm for obtaining the consistent result and propose the node-splitting solution which is described in detail in Sect. 3.4. TNA-SL takes directed trust edges between pairs as input, and can be used to derive a level of trust between arbitrary parties that are interconnected through the network.

TNA-SL therefore has a general applicability and is suitable for many types of trust networks.

Trust is a concept that crosses disciplines and also domains. The focus of different definitions differs on the basis of the goal and the scope of the works. Two generalized definitions of trust were defined by Jøsang et al. (2007), which he called *reliability trust* (the term "evaluation trust" is more widely used by the other researchers, therefore we use this term) and *decision trust* respectively will be used for this work. Evaluation trust can be interpreted as the reliability of something or somebody. It can be defined as the subjective probability by which an individual, $A$, expects that another individual, $B$, performs a given action on which their welfare depends. On the other hand, the decision trust captures a broader concept of trust. It can be defined as the extent to which one party is willing to depend on something or somebody in a given situation with a feeling of relative security, even though negative consequences are possible. Trust is a belief which can be considered as a measurable continuous subjective quantity (Lesani and Montazeri 2009). The trust value has a lower bound which means no trust and an upper bound which refers to complete trust and partial trust comes in between. Taking 0 as lower and 1 as upper bounds, trust results will be in the range of [0,1]. If the trust relationship in a network is abstracted to a trust graph, to infer personalized trust from one node to another, algorithms are needed to find and select from the paths that exist from the first node to second. Aggregation methods are also required to combine trust values when the values are obtained from several nodes. We have applied the belief combination operators to combine trust. We have proposed our DSPG algorithm to represent the sub-graph comprised of some selected paths between a source and a sink as a closed form of serial and parallel connections. Here, closed form refers to a subset of nodes adjunct to the path from source to sink or target. The trust from the source to the sink is then inferred from the obtained closed form using the Subjective Logic *discounting* and *cumulative fusion* operators for serial and parallel belief combination, respectively. The correspondences of these operators with binary logic or set operators are presented in Table 3.2.

**Discounting** is used to compute trust along a chain of trust edges. Assume agent $A$ has referral trust in another agent $B$, denoted by $\omega_B^A$, for the purpose of judging the functional or referral trustworthiness of $x$, and $B$ has functional or referral trust in $x$, denoted by $\omega_x^B$. Agent $A$ can then derive her trust in $x$ by discounting $B$'s trust in $x$ with $A$'s trust in $B$, denoted by $\omega_x^{A:B}$. If $A$'s trust in $B$ on $x$ is $\omega_B^A = (b_B^A, d_B^A, u_B^A, a_B^A)$ opinion and $B$'s opinion about $x$ is $\omega_x^B = (b_x^B, d_x^B, u_x^B, a_x^B)$ then to infer $A$'s

**Table 3.2** Correspondence between subjective logic and binary logic or set operators for serial and parallel trust calculation

| Subjective logic operator | Symbol | Binary logic/set operator | Symbol | Subjective logic notation |
|---|---|---|---|---|
| Discounting | $\otimes$ | Transitivity | : | $\omega_x^{A:B} = \omega_B^A \otimes \omega_x^B$ |
| Cumulative fusion | $\oplus$ | Consensus | $\diamond$ | $\omega_x^{A\diamond B} = \omega_x^A \oplus \omega_x^B$ |

opinion of $x$, $B$'s opinion about $x$ is reduced with $A$'s opinion of $B$. The discounting operator is defined as follows:

$$\omega_x^{A:B} = \omega_B^A \otimes \omega_x^B = \begin{cases} b_x^{A:B} = b_B^A b_x^B \\ d_x^{A:B} = b_B^A d_x^B \\ u_x^{A:B} = d_B^A + u_B^A + b_B^A u_x^B \\ a_x^{A:B} = a_x^B \end{cases} \tag{3.6}$$

Here, the superscript denotes the trusting and the subscript denotes the trusted. If only belief element is present, i.e. $\omega_B^A = (b_B^A, 0, 1 - b_B^A, a_B^A)$ and $\omega_x^B = (b_x^B, 0, 1 - b_x^B, a_x^B)$ then the operator can be simplified to the Eq. (3.7).

$$\omega_x^{A:B} = \omega_B^A \otimes \omega_x^B = \begin{cases} b_x^{A:B} = b_B^A b_x^B \\ d_x^{A:B} = 0 \\ u_x^{A:B} = u_B^A + b_B^A u_x^B \\ a_x^{A:B} = a_x^B \end{cases} \tag{3.7}$$

The effect of the discounting operator is to increase uncertainty and decrease belief and disbelief in the transitive chain.

**Cumulative fusion** between two opinions is an opinion that reflects both of the initial opinions fairly and equally.

If $A$ and $B$'s believe in $x$ with corresponding beliefs of $\omega_x^A = (b_x^A, d_x^A, u_x^A, a_x^A)$ and $\omega_x^B = (b_x^B, d_x^B, u_x^B, a_x^B)$, assuming $a_x^A = a_x^B$, then the cumulative fusion operator is defined as below:

$$\omega_x^{A \Diamond B} = \omega_B^A \oplus \omega_x^B = \begin{cases} b_x^{A \Diamond B} = (b_x^A u_x^B + b_x^B u_x^A)/(u_x^A + u_x^B - u_x^A u_x^B) \\ d_x^{A \Diamond B} = (d_x^A u_x^B + d_x^B u_x^A)/(u_x^A + u_x^B - u_x^A u_x^B) \\ u_x^{A \Diamond B} = (u_x^A u_x^B)/(u_x^A + u_x^B - u_x^A u_x^B) \\ a_x^{A \Diamond B} = a_x^A. \end{cases} \tag{3.8}$$

If only belief element is present, i.e. $\omega_B^A = (b_B^A, 0, 1 - b_B^A, a_B^A)$ and $\omega_x^B = (b_x^B, 0, 1 - b_x^B, a_x^B)$ then the operator can be simplified to

$$\omega_x^{A \Diamond B} = \omega_B^A \oplus \omega_x^B = \begin{cases} b_x^{A \Diamond B} = (b_x^A + b_x^B - 2b_x^A b_x^B)/(1 - b_x^A b_x^B) \\ d_x^{A \Diamond B} = 0 \\ u_x^{A \Diamond B} = (u_x^A u_x^B)/(u_x^A + u_x^B - u_x^A u_x^B) \\ a_x^{A \Diamond B} = a_x^A. \end{cases} \tag{3.9}$$

The effect of the cumulative fusion operator is to increase belief and disbelief and decrease uncertainty.

**Example Derivation of Trust Measures:** The *discounting* and *cumulative fusion* operators will be used for the purpose of deriving trust measures applied to the trust graph of Figs. 3.4 and 3.5.

In the case of Fig. 3.4, the edge trust values will all be set equal as:

$$\omega_B^A = \omega_C^B = \omega_D^C = (0.9, 0.0, 0.1, 0.5)$$

By applying the *discounting* operator to the expression of Eq. (3.7), the derived trust value evaluates to:

$$\omega_D^{A:B:C} = \omega_B^A \otimes \omega_C^B \otimes \omega_D^C = (0.729, 0.000, 0.271, 0.5)$$

In the case of Fig. 3.5, the edge trust values will all be set equal as:

$$\omega_B^A = \omega_D^B = \omega_C^A = \omega_D^C = (0.9, 0.0, 0.1, 0.5)$$

By applying the *discounting* and *cumulative fusion* operators to the expression of Eq. (3.9), the derived indirect trust measure can be computed. The expression for the derived trust measure and the numerical result is given below.

$$\omega_D^A = (\omega_B^A \otimes \omega_D^B) \otimes (\omega_C^A \otimes \omega_D^C)$$
$$= (0.895, 0.000, 0.105, 0.5)$$

The proposed algorithm has two major steps. In the first step, in a given directed graph, it finds all possible paths from a source to a target node by using a normal depth-first search algorithm. In the second step, it calculates the trust value from source to target node by using the formula (3.6) for a single path and by using the formula (3.8) for multiple paths to combine all possible paths within a preset threshold value.

---

**Algorithm 3.1** *DFSFindPaths(source, target, graph)*

---

**Input:** *source* is the source node and *target* is the target node of a link in the trust graph *graph*
**Output:** *Paths* include all the paths from *source* to *target*
**Begin**
1. *arcs* : = { }//initialise a temporary variable *arcs*
2. **for** each *arc* in *graph*
3.   **if** *arc.source* == *source* **and** *arc.target* is not in *Path*
     **then** *arcs* : = *arcs* ∪ {*arc*}//
     //end-for
4. **for** each *arc* **in** *arcs* **do**
5. **if** *arc.target* == *target* and *arc.variant* == *'functional'*
     //add the arc to *Path*, and add the Path to *Paths*
6. **then** *paths* : = *paths* ∪ {*Path* ∪ {*arc*}}
7. **else if** *arc.target* ≠ *target* and *arc.variant* == *'referral'*
8.   **then** {*path* : = *path* ∪ {*arc*}//add the arc to Path
9.     *DFSFindPaths(arc.target, target, graph)*

---

(continued)

(continued)

---

**Algorithm 3.1** *DFSFindPaths*(*source, target, graph*)

---

10.   *path* : = *path* \{*arc*}//remove the arc from the current path
    }//end-if
    //end-if
    //end-for
    **End**

---

In the above algorithm which is to find all the paths from one source node to a target node, arc.source and arc.target stand for the source node and the target node of a link in a graph, arc.variant represents the type of trust on that link, arc.variant = "functional" means that the trust on that link is a functional trust and arc.variant = "referral" means that the trust on that link is a referral trust. A simplified algorithm is given below, for more details; please refer to (Jøsang et al. 2006). The following algorithm is a recursive algorithm. In order to make the algorithm work, we need two global variables, *Paths* and *Path*, *Paths* is a set of paths, and *Path* is a set of arcs representing a path. At the end of executing the algorithm, *Paths* will contain all the paths from the specified source to the specified target, and at any time during the execution, *Path* gives the current working path. Initially, both variables are empty sets.

## 3.4 Trust Path Dependency and Network Simplification

A potential limitation with the TNA-SL is that complex trust networks must be simplified to *series–parallel* networks in order for TNA-SL to produce consistent results. The simplification consisted of gradually removing the least certain trust paths until the whole network can be represented in a series–parallel form. As this process removes information, it is intuitively sub-optimal. In this section, we describe how TNA-SL can preserve consistency without removing information. Inconsistency can result from dependence between separate trust paths, which when combined will take the same information into account several times. Including the same trust edges multiple times will, by definition, produce an inconsistent result. In this section, we propose an Optimal TNA-SL which avoids this problem by allowing the trust measure of a given trust edge to be split into several independent parts, so that each part is taken into account by separate trust paths. The result of this approach was compared with the analysis based on networks simplification and achieved the same result.

We have used basic constructs of directed graphs to represent transitive trust networks for this part of work, and have added some notation elements which allowed us to express trust networks in a structured way. A single trust relationship can be expressed as a directed edge between two nodes in a trust network that

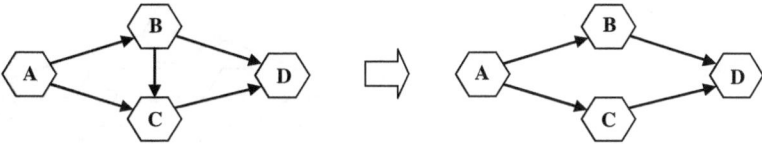

**Fig. 3.6** Network simplification by removing the weakest path

represents the trust source and the trust target of that edge. For example, the edge [*A*, *B*] means that *A* trusts *B* which is equivalent to subjective logic opinion representation $\omega_B^A$. The symbol ":" is used to denote the transitive connection of two consecutive trust edges to form a transitive trust path. The trust relationships of Fig. 3.4 can be expressed as:

$$([A, D]) = ([A, B] : [B, C] : [C, D]) \tag{3.10}$$

Let us now turn to the combination of parallel trust paths, as illustrated in Fig. 3.5. As presented in Table 3.2, we have used the symbol "$\diamond$" to denote the graph connector for this purpose. The "$\diamond$" symbol visually resembles a simple graph of two parallel paths between a pair of agents, so it is natural to use it for this purpose. In short notation, a combination of the two parallel trust paths from *A* to *D* in Fig. 3.5 is then expressed as:

$$([A, D]) = (([A, B] : [B, D])) ([A, C] : [C, D]) \tag{3.11}$$

It can be noted that Fig. 3.5 contains two parallel paths.

Trust networks can have dependent paths. This can be illustrated on the left-hand side of the example in Fig. 3.6.

The expression for the graph on the left-hand side of Fig. 3.6 would be:

$$([A, D]) = (([A, B] : [B, D]) \diamond ([A, C] : [C, D]) \diamond ([A, B] : [B, C] : [C, D])) \tag{3.12}$$

A problem with Eq. (3.12) is that the arcs [*A,B*] and [*C,D*] appear twice, and the expression is therefore not canonical. Trust network analysis with subjective logic may produce inconsistent results when applied directly to non-canonical expressions. It is therefore desirable to express graphs in a form where an arc only appears once.

A method for canonicalization based on network simplification was described in (Jøsang et al. 2006). Simplification consists of removing the weakest, i.e. the least certain paths, until the network becomes a directed series–parallel network which can be expressed in a canonical form. Assuming that the path ([*A,B*]:[*B,C*]:[*C,D*]) is the weakest path in the graph on the left-hand side of Fig. 3.6, network simplification of the dependent graph would be to remove the edge [*B*, *C*] from the graph, as illustrated on the right-hand side of Fig. 3.6. Since the simplified graph is equal to that of Fig. 3.5, the formal expression is the same as Eq. (3.11).

**Fig. 3.7** Node splitting of a
trust network to produce
independent paths

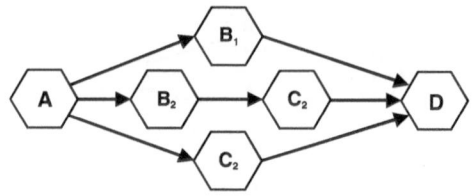

The existence of a dependent edge in a graph is recognized by multiple instances of the same edge in the trust network expression. Node splitting is a new approach to achieving independent trust edges. This is achieved by splitting the target edge of a given dependent edge into as many different nodes as there are different instances of the same edge in the exploded notation. A general directed trust graph is based on directed trust edges between pairs of nodes. It is desirable not to put any restrictions on the possible trust arcs except that they should not be cyclic. This means that the set of possible trust paths from a given source $X$ to a given target $Y$ can contain dependent paths. The left-hand side of Fig. 3.6 shows an example of a trust network with dependent paths.

In the non-canonical expression for the left-hand side of the trust network of Fig. 3.6:

$$([A, D]) = ([A, B] : [B, D]) \Diamond ([A, C] : [C, D]) \Diamond ([A, B] : [B, C] : [C, D])) \quad (3.13)$$

The edges $[A, B]$ and $[C, D]$ appear twice. Node splitting in this example consists of splitting the node $B$ into $B_1$ and $B_2$, and the node $C$ into $C_1$ and $C_2$. This produces the trust network with independent paths in Fig. 3.7 with canonical expression:

$$([A, D]) = ([A, B_1] : [B_1, D]) \Diamond ([A, C_1] : [C_1, D]) \Diamond ([A, B_2] : [B_2, C_2] : [C_2, D]))$$
$$(3.14)$$

Node splitting must be translated into opinion splitting in order to apply subjective logic. The principle for opinion splitting will be to separate the opinion on the dependent edge into two independent opinions that when cumulatively fused produce the original opinion. This can be called fission of opinions, and will depend on a fission factor ø that determines the proportion of evidence assigned to each independent opinion part. The mapping of an opinion $\omega = (b, d, u, a)$ to Beta evidence parameters Beta$(r, s, a)$ and linear splitting into two parts Beta$(r_1, s_1, a_1)$ and Beta$(r_2, s_2, a_2)$ as a function of the fission factor ø is:

$$\text{Beta}(r_1, s_1, a_1) : \begin{cases} r_1 = \dfrac{\phi 2b}{u} \\[2mm] s_1 = \dfrac{\phi 2d}{u} \\[2mm] a_1 = a \end{cases} \quad \text{Beta}(r_2, s_2, a_2) : \begin{cases} r_2 = \dfrac{(1 - \phi) 2b}{u} \\[2mm] s_2 = \dfrac{(1 - \phi) 2d}{u} \\[2mm] a_2 = a \end{cases} \quad (3.15)$$

The reverse mapping of these evidence parameters into two separate opinions will produce:

$$
\omega_1 : \begin{cases} b_1 = \dfrac{\phi b}{\phi(b+d)+u} \\[2mm] d_1 = \dfrac{\phi d}{\phi(b+d)+u} \\[2mm] u_1 = \dfrac{u}{\phi(b+d)+u} \\[2mm] a_1 = a \end{cases}
\qquad
\omega_2 : \begin{cases} b_2 = \dfrac{(1-\phi)b}{(1-\phi)(b+d)+u} \\[2mm] d_2 = \dfrac{(1-\phi)d}{(1-\phi)(b+d)+u} \\[2mm] u_2 = \dfrac{u}{(1-\phi)(b+d)+u} \\[2mm] a_2 = a \end{cases}
\qquad (3.16)
$$

It can be verified that $\omega_1 \oplus \omega_2 = \omega$, as expected.

When deriving trust values from the exploded trust network of Eq. (3.12) we are interested in knowing its certainty level as compared with a simplified network. We are interested in the expression for the uncertainty of $\omega_D^A$ corresponding to the trust expression of Eq. (3.12). Since the node splitting introduces parameters for splitting opinions, the uncertainty will be a function of these parameters. By using Eq. (3.6) the expressions for the uncertainty in the trust paths of Eq. (3.14) can be derived as:

$$
\begin{aligned}
u_D^{A:B_1} &= d_{B_1}^A + u_{B_1}^A + b_{B_1}^A u_D^{B_1} \\
u_D^{A:C_1} &= d_{C_1}^A + u_{C_1}^A + b_{C_1}^A u_D^{C_1} \\
u_D^{A:B_2:C_2} &= b_{B_2}^A d_{C_2}^{B_2} + d_{B_2}^A + u_{B_2}^A + b_{B_2}^A u_D^{B_2} + b_{B_2}^A b_{C_2}^{B_2} u_D^{C_2}
\end{aligned}
\qquad (3.17)
$$

By using Eqs. (3.6) and (3.17), the expression for the uncertainty in the trust network of Eq. (3.14) can be derived as:

$$
u_D^A = \frac{u_D^{A:B_1} u_D^{A:C_1} u_D^{A:B_2:C_2}}{u_D^{A:B_1} u_D^{A:C_1} + u_D^{A:B_1} u_D^{A:B_2:C_2} + u_D^{A:C_1} u_D^{A:B_2:C_2} - 2u_D^{A:B_1} u_D^{A:C_1} u_D^{A:B_2:C_2}}
\qquad (3.18)
$$

By using Eqs. (3.16), (3.17) and (3.18), the uncertainty value of the derived trust $\omega_D^A$ according to the node splitting principle can be computed. This value depends on the trust edge opinions and on the two splitting parameters $\phi_B^A$ and $\phi_D^C$. By fixing the opinion values as in the example of Eq. (3.19) as below:

$$
\omega_B^A = \omega_D^B = \omega_C^A = \omega_D^C = \omega_C^B = (0.9,\ 0.0,\ 0.1,\ 0.5) \qquad (3.19)
$$

The computed trust values for the two possible simplified graphs are:

$$
(\omega_B^A \otimes \omega_D^B) \oplus (\omega_C^A \otimes \omega_D^C) = (0.895,\ 0.0,\ 0.105,\ 0.5) \qquad (3.20)
$$

$$
\omega_B^A \otimes \omega_C^B \otimes \omega_D^C = (0.729,\ 0.0,\ 0.271,\ 0.5) \qquad (3.21)
$$

The conclusion which can be drawn from this is that the optimal values for the splitting parameters are $\phi_B^A = \phi_D^C = 1$ because that is when the uncertainty is at its lowest. In fact the uncertainty can be evaluated to $u_D^A = 0.105$ in that case, which is equal to the uncertainty of Eq. (3.20). This is equivalent to the case of trust network simplification where the edge $[B, C]$ is removed from the left-hand side graph of Fig. 3.6. The least optimal values for the splitting parameters are when $\phi_B^A = \phi_D^C = 0$, resulting in $u_D^A = 0.271$ which is equal to the uncertainty of Eq. (3.21). This is thus equivalent to the absurd trust network simplification where the edges are $[A, C]$ and $[B, D]$, and thereby the most certain trust paths are removed from the left-hand side graph of Fig. 3.6. Given the edge opinion values used in this example, $([A,B]:[B,C]:[C, D])$ is the least certain path of the left-hand side graph of Fig. 3.6. It turns out that the optimal splitting parameters for analyzing the right-hand side graph of Fig. 3.6 produces the same result as network simplification when this particular least certain path is removed.

## 3.5 Related Work

Golbeck (2005) has proposed a trust inference algorithm named *TidalTrust* which performs the inference through the strongest of the shortest paths. It infers trust incrementally from the sink, level by level back to the source. For each node, trust to the sink is computed locally from the information available from its neighbors. Computed trust values for nodes in the current level are used in computing trust values for nodes in the level that is one level farther from the sink. Massa's (Massa and Avesani 2007) *MoleTrust* also follows similar technique to *TidalTrust*. The major difference is in the set of trusted raters' considered. Both *TidalTrust* and *MoleTrust* perform a breadth-first search of the trust network. *TidalTrust* considers all raters at the minimum depth by finding the shortest path distance from the active user but *MoleTrust* considers all raters up to a specified maximum depth. The name for the algorithm *TidalTrust* was chosen because calculations sweep forward from source to sink in the network, and then pull back from the sink to return the final value to the source. The source node begins a search for the sink. It will poll each of its neighbors to obtain their rating of the sink. Each neighbor repeats this process, keeping track of the current depth from the source. Each node will also keep track of the strength of the path to it. Nodes adjacent to the source will record the source's rating assigned to them. Each of those nodes will poll their neighbors. The strength of the path to each neighbor is the minimum of the source's rating of the node and the node's rating of its neighbor. The neighbor records the maximum strength path leading to it. Once a path is found from the source to the sink, the depth is set at the maximum depth allowable. Since the search is proceeding in a BFS fashion, the first path found will be at the minimum depth. The search will continue to find any other paths at the minimum depth. Once this search is complete, the trust threshold is established by taking the maximum of the trust paths leading to the sink. With the *max* value established, each node can complete the calculations of a weighted average by taking information from nodes that they have rated at or above the *max* threshold.

$$t_{is} = \frac{\sum_{(j \in adj(j)|t_{ij} \geq \text{ max})} t_{ij} t_{js}}{\sum_{(j \in adj(j)|t_{ij} \geq \text{ max})} t_{ij}} \tag{3.22}$$

## 3.6  Chapter Summary

This chapter introduces the proposed *DSPG* inference model for inferring trust relationships using subjective logic. The proposed model can be used for inferring a trust value from one node to another where no direct connection exists between the nodes in a trust network. Trust Inference can be used to recommend to one node how much to trust another node or what trust rating one person may give another unknown person if there is any indirect link present in the network. Thus the trust inference can be used to find more nodes as neighbors who are indirectly connected, to improve recommendation quality. The definitions and notations are presented followed by the proposed trust inference model. Finally, the popular trust inference algorithm *TidalTrust* has described.

# Chapter 4
# Online Survey on Trust and Interest Similarity

**Abstract** A remarkable growth in quantity and popularity of online social networks has been observed in recent years. There is a good number of online social networks existing which have over 100 million registered users. Many of these popular social networks offer automated recommendations to their users. These automated recommendations are normally generated using collaborative filtering systems based on the past ratings or opinions of the similar users.

**Keywords** Recommender systems · Trust · Inference · User tag · Web 2.0 · Social networks

A remarkable growth in quantity and popularity of online social networks has been observed in recent years. There is a good number of online social networks existing which have over 100 million registered users. Many of these popular social networks offer automated recommendations to their users. These automated recommendations are normally generated using collaborative filtering systems based on the past ratings or opinions of the similar users. Alternatively, trust among the users in the network also can be used to find the neighbors while making recommendations. To obtain an optimum result, there must be a positive correlation existing between trust and interest similarity. Although a positive relationship between trust and interest similarity is assumed and adopted by many researchers, no survey work on real life people's opinions to support this hypothesis was found. This chapter presents the result of the survey on the relationship between trust andinterest similarity. The result supports the assumed hypothesis of a positive relationship between the trust and interest similarity of the users.

## 4.1 Related Work

Trust-aware recommender systems have gained the attention of many researchers in recent years, where, instead of the most similar users' opinions, the most trusted users' opinions are considered to make automated recommendations. The well

T. Bhuiyan, *Trust for Intelligent Recommendation*,                                                    53
SpringerBriefs in Electrical and Computer Engineering,
DOI: 10.1007/978-1-4614-6895-0_4, © The Author(s) 2013

known reviewers' community Epinions (www.epinions.com) provides information filtering facilities based upon a personalized web of trust and it is stated that the trust-based filtering approach has been greatly approved and appreciated by Epinions' members (Guha et al. 2004). The sociology and social psychology researchers address factors associated with trust in many ways, but, in the existing literature, not much work has been found which directly addresses how trust relates to similarity. Abdul-Rahman and Hailes (2000) claim that, given some predefined domain and context, people create ties of friendship and trust primarily with people with profiles that resemble their own profile of interest. Research in social psychology offers some important results for investigation of interactions between trust and similarity. However, most relevant studies primarily focus on interpersonal attractions rather than trust and its possible coupling with similarity (Ziegler and Golbeck 2007). A positive relationship between attitude similarity and friendship has been shown by Burgess and Wallin (1943) in their work about homogamy of social attributes with respect to engaged couples; similarity could be established for nearly every characteristic examined. However, according to Berscheid (1998), these findings do not justify conclusions about positive effects of similarity on interpersonal attraction by themselves. The first large-scale experimental studies were conducted by Newcomb and Byrne. Newcomb (1961) focused on friendships between American college students and his work became one of the most popular methods to describe friendship formation. His work also confirmed that attitude similarity is a determinant of attraction. By means of his longitudinal study, Newcomb could reveal a positive association between attraction and attitudinal value similarity. Byrne (1961, 1971) performed extensive research and experiments with large-scale data in the area of attraction, applying the famous bogus stranger technique. The results of Byrne's experiment aligned well with Newcomb's findings and confirmed that attitude similarity is a determinant of attraction. Social psychologists have identified some other likely factors accounting for the similarity-attraction associations. For example, the information that another person possesses similar attitudes may suggest his sympathy towards the individual. Snyder and Fromkin (1980) reveal that perceiving very high similarity with another individual may even evoke negative sentiments towards that person. Jensen (Jensen et al. 2002) assumed similarity as a strong predictor of friendship. Evidence from socio-psychological research thus gives hints proposing the existence of positive interactions between trust and interest similarity. Montaner (Montaner et al. 2002) claims that trust should be derived from user similarity, implying that friends are those people with very similar natures. Golbeck (2009) has shown the potential implications for using trust in user interfaces in the area of online social network.

Ziegler and Lausen (2004) mention that in order to provide meaningful results, trust must reflect user similarity to some extent because recommendations only make sense when obtained from like-minded people exhibiting similar taste. They provide empirical results obtained from one real, operational community and verify their hypothesis for the domain of book recommendations. Later on, Ziegler and Golbeck (2007) have proposed a framework which suggests that a positive co-relationship exists between trust and interest similarity which means "the more similar two people, the greater the trust between them". They show the relationship between trust and overall similarity assuming that given an application domain, people's trusted peers are on average considerably more similar to their sources of trust than arbitrary peers. They experimentally prove that there exists a significant correlation between the trust expressed by the users and their profile similarity based on the recommendations they made in the system. This correlation is further studied as survey-based experiments by Golbeck (2006). They proposed that if A denotes the set of all community members, trust $(a_i)$ the set of all users trusted by $a_i$, and $sim$ A $\times$ A $\rightarrow$ [−1, +1] some similarity function:

$$\sum_{a_i \in A} \frac{\sum_{a_j \in trust(a_i)} sim(a_i, a_j)}{|trust(a_i)|} \gg \sum_{a_i \in A} \frac{\sum_{a_j \in A\{a_i\}} sim(a_i, a_j)}{|A| - 1} \qquad (4.1)$$

By using movie rating data in their experiment, they have shown that as the trust between users' increases, the difference in the ratings they assign decreases. It indicates that a positive correlation exists between trust and interest similarity among the users of the networks. Our survey results also support these findings which are discussed in detail in the next sections.

## 4.2 Survey on the Relationship Between Interest Similarity and Trust

### 4.2.1 Study Objective

The major objective of this survey is to collect information about the user's view regarding the relationship between trust and interest similarity. We set the questioner to obtain information about three main sub-topics listed below:

- Acceptance of Online Recommendation
- Perceptions about Other Online Users
- Relationship between Trust and Interest Similarity.

The list of questions asked to the respondent is included in Appendix A.

### 4.2.2 Study Design

An online survey methodology was chosen in order to maximize the geographical spread of respondents, speed of data collection and anonymity of participants (Harding and Peel 2007; Peel 2009). The survey was designed by using www.SurveyMonkey.com and it contained 10 questions. The questions were developed based on key issues in the academic and lay literatures and experiential

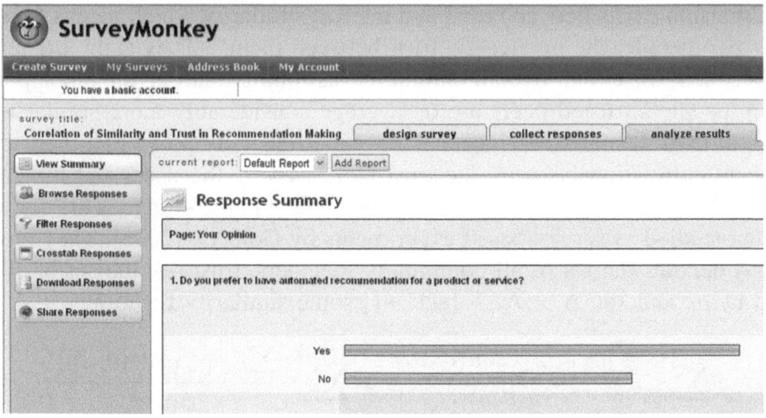

**Fig. 4.1** A screen shot of the survey tool

knowledge. In creating a survey, Coughlan et al. (2009) states that the investigator only should ask what is necessary and not what might be interesting. Trying to answer too many things usually means none of them are answered well. For this reason, the questions were kept to a minimum number. It was stated in the introductory information that the study focuses on automated recommendation particularly in the online environment. It also stated that "you will remain anonymous and any identifiable information you provide will be changed. Information you provide will be held on Survey Monkey's server, however, Survey Monkey guarantees that the data will be kept private and confidential" (Fig. 4.1).

The researcher's contact information was provided for respondents to ask any questions about the study before deciding whether to take part, and information about further sources of support and information were provided. Respondents were free to exit the survey at any point without giving reason and a response was not mandatory for all the questions asked. Australian Psychological Society Ethical Guidelines were adhered to and Queensland University of Technology Human Research Ethics Committee granted ethical approval (Approval number 0900001051 granted on 19/10/2009 in the category of "Human non-HREC"). Czaja and Blair (2005) state that an effective survey has three essential characteristics: it is a valid measure of the factors of interest, it convinces the respondents to cooperate, and it elicits acceptably accurate information (Czaja and Blair 2005). All these three characteristics were kept in mind while designing the survey. The survey was piloted and refined before going live.

### 4.2.3 Recruitment and Data Collection

Respondents were recruited using strategic opportunistic sampling. Five recruiting eMails were sent to QUT HDR eMail list, University Alumni Association and personal contacts. The study was also publicized through the social network Facebook in Australia, UK and USA regions. Data were collected between November 2009 and February 2010 with the majority of responses occurring within the 1 month of the study being publicized. Due to the lack of available time; we had to restrict the survey to 4 months only. The time limitation of this survey also limits the number of respondents. A little longer time could help to increase the number of participants of the survey (Innovation Network 2009).

### 4.2.4 Respondents

A total of 408 respondents, from different parts of the world including Australia, UK, USA, Bangladesh and China, participated in the study conducted online. Though there was no age limit specified for the survey, the invitation email to participate in the online survey was sent to adult online users only who were at least 18 years old. Respondents included both male and female online users of different age groups.

### 4.2.5 Results

We received 408 persons in total as the respondents of our online survey through www.SurveyMonkey.com within the allocated 4 months time period. As the number of questions was limited to only 10, all of the participants answered all questions without skipping a single one. As the objective of the study; we categorized our findings in three different sections which are discussed in the following sub-sections below:

#### 4.2.5.1 Acceptance of Online Recommendation

Among 408 participants, 58 % of respondents express their positive opinions about online recommendation. We asked direct questions like "do you prefer to have automated recommendation for a product or service?" (Fig. 4.2).

**Fig. 4.2** Acceptance of online recommendation

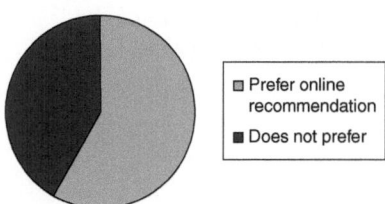

**Fig. 4.3** Recommendation
source

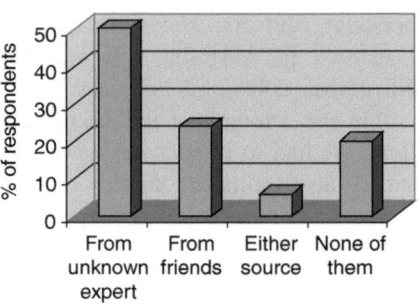

We also asked indirect questions like "assume that an unknown automobile
expert *A* and one of your friends *B* who is not an expert about cars is available for
recommendation when you are going to buy a car; which recommendation will you
prefer?" (Fig. 4.3). 50 % of respondents prefer the expert opinion and 24 % prefer
that the recommendation came from their friends they know personally. 6 % of
respondents do not have any concern about the source of the recommendation,
they are happy to receive recommendation from either source. Only 20 % do not
appreciate any recommendations irrespective of the source of recommender.

### 4.2.5.2 Perceptions About Other Online Users

A little more than half of the total respondents (52.5 %) consider some of them as
a friend to whom they met online and others found it difficult to trust them as a
friend (Fig. 4.4). About 48 % of people think it is unnecessary to rate their online
friends as to how much they trust them.

Only 28 % of people think it would be helpful to mention how much they trust
their online friend. While another 13 % would not bother about it and 11 % are
undecided. Even if they accept online mates as friends; most of them (59 %) do
not bother about rating online friends.

### 4.2.5.3 Relationship Between Trust and Interest Similarity

Forty percentage of the respondents express their direct positive opinion about the
relationship between trust and interest similarity. A large portion (29 %) of

**Fig. 4.4** Perceptions about
the people they met online

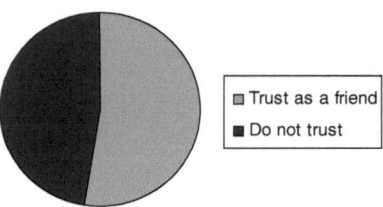

respondents expresses their uncertainty about the issue; which could be due to the lack of understanding about the meaning of interest similarity. From the informal feedback of the respondents it has been discovered that many of them were little confused about the interpretation of interest similarity. However, only 31 % expressed that they do not find any relationship between trust and interest similarity.

Given the choice between a recommendation from a friend with similar tastes and a friend with different likes; it was pre-assumed that most people will choose the recommendation from a friend with similar tastes. Our result shows that 66.7 % of users prefer a recommendation from a friend with similar tastes rather than a friend with different tastes. Here, the taste should be limited to a particular scope or domain such as movies, books or holiday destination recommendations. It is reflected from the result here that for a given domain, people prefer their source of recommendations to be a friend with similar tastes. The scope or domain limitation is important for this opinion.

## 4.2.6  Findings and Discussion

Several issues have been discovered during the survey. In previous work Sinha and Swearingen (2001) presented their findings that people prefer to accept recommendation from their family and friends rather than auto generated recommendation. Our finding indicates that this view has changed in the last decade. Our results also indicate that most of the people are unsure about considering other online users as friends and are not interested in rating them (Fig. 4.5). However, the overall attitude of the online user about the relationship between trust and interest similarity is positive which supports the basic hypothesis of our survey. The findings are discussed below.

### 4.2.6.1  People Start Relying on the Online Recommendations

Unlike the findings of Sinha and Swearingen (2001); who claim people prefer receiving recommendations from people they know and trust, such as from their

**Fig. 4.5** Overall view about other online users

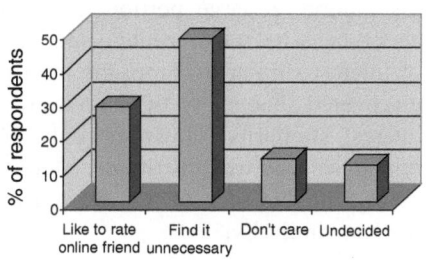

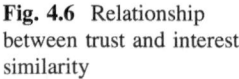

**Fig. 4.6** Relationship
between trust and interest
similarity

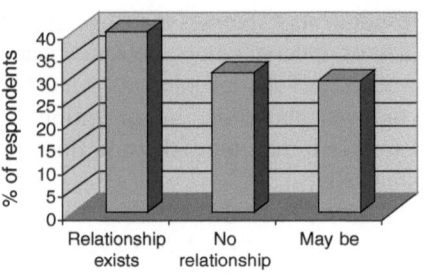

family members or friends rather than from recommender systems; our result shows that people prefer to rely on expert opinion irrespective of whether the source is known or trusted as long as it comes from an expert (Fig. 4.6).

### 4.2.6.2 People are Unsure Whether to Consider Other Online Users as Friends

Many people consider the other users they meet online as friends but almost the same number of people thinks the opposite. It is found that they are uncertain about treating the people they meet online as their friends.

### 4.2.6.3 A Positive Relationship Exists Between Trust and Interest Similarity

Most of the people believe a positive relationship exists between trust and interest similarity among different users. They prefer to trust more the opinions of those with similar tests in a particular matter. On a given scope or topic, people generally like to accept the opinion or recommendation from similar or like minded people. It gives them confidence that the recommendation is came from a similar minded user so there is a more possibility that they are going to like it. This is reflected on the findings above.

From the Fig. 4.6, it can be noticed that 40 % of the respondents express their direct positive opinion about the relationship between trust and interest similarity. These respondents expressed their positive opinion directly in favor of our assumption. A large portion (29 %) of respondents expresses their uncertainty about the issue; which could be due to the lack of understanding about the meaning of interest similarity. From the informal feedback of the respondents it has been discovered that many of them were little confused about the interpretation of interest similarity. However, only 31 % expressed that they do not find any relationship between trust and interest similarity. In other words, total 69 % disagree with the idea of non-existence of positive relationship between trust and interest similarity.

### *4.2.7  Conclusions*

The positive correlation between trust and interest similarity has been assumed for a long time in the area of recommender systems. Our survey results support that hypothesis strongly. It is also found that in general, people prefer online recommendation if it comes from a subject expert. But people are uncertain about accepting the people they met online as their friends. We believe that our findings from this survey will have a great impact in the area of recommender system research; especially where discovering user interest similarity plays an important role. The findings of the existence of positive correlation between trust and interest similarity gives us the confidence that the interest similarity value can be used as trust data to build a trust network in the absence of direct explicit trust.

## 4.3  Chapter Summary

This chapter describes the details of the survey conducted for investigating the opinions' of online users regarding the relationship between trust and interest similarity. It has been found that the survey result supports the assumption of positive correlation between trust and interest similarity. This finding is used as a basis for utilizing interest similarity value as a trust value in the absence of explicit trust rating data. The positive correlation between trust and interest similarity has been assumed for a long time in the area of recommender systems. Our survey result strongly supports that hypothesis. It is also found that in general, people prefer online recommendation if it comes from a subject expert. But people are uncertain about accepting the people they met online as their friends. We believe that our findings from this survey will have a great impact in the area of recommender system research.

# Chapter 5
# *SimTrust*: The Algorithm for Similarity-Based Trust Network Generation

**Abstract** The essential recommendation making mechanism of current recommender systems is to firstly identify the target user's neighbors based on user profile similarity, and then suggest the target user items that the neighbors have liked in the past. One weakness of the current recommender systems is the inaccurate neighborhood identification due to insufficient user profile data and limited understanding to users' interests. Trust can be used for neighborhood formation to generate automated recommendation. User-assigned explicit trust rating, such as how much they trust each other, can be used for this purpose. However, reliable explicit trust data is rarely available.

**Keywords** Recommender systems · Trust · Inference · User tag · Web 2.0 · Social networks

The essential recommendation making mechanism of current recommender systems is to firstly identify the target user's neighbors based on user profile similarity, and then suggest the target user items that the neighbors have liked in the past. One weakness of the current recommender systems is the inaccurate neighborhood identification due to insufficient user profile data and limited understanding to users' interests. Trust can be used for neighborhood formation to generate automated recommendation. User-assigned explicit trust rating, such as how much they trust each other, can be used for this purpose. However, reliable explicit trust data is rarely available.

The specific technology of new generation Internet programs known as Web 2.0 has created a more rich and complex online phenomenon, enabling self-expression, communication, and user interaction online. Instead of the Internet being an enormous reading resource repository, the new generation of the Web becomes a platform for users to participate, collaborate and interact online. In this new generation of read-and-write Web, the *User Generated Content* (UGC) provided by Web users, e.g. in the form of category tags (i.e. user-defined labels for organizing their items), blogs, reviews and comments becomes an important information resource in addition to traditional website materials. The richness of

T. Bhuiyan, *Trust for Intelligent Recommendation*,
SpringerBriefs in Electrical and Computer Engineering,
DOI: 10.1007/978-1-4614-6895-0_5, © The Author(s) 2013

new UGC sites challenges the current personalization techniques and provides new possibilities for accurately profiling users. Thus, how to incorporate the new features and practices of Web 2.0 into personalization applications becomes an important and urgent research topic. In this chapter we propose a new method of developing trust networks based on user's interest similarity in the absence of explicit trust data. To identify the interest similarity, we have used users' personalized tagging information. This trust network can be used to find the neighbors to make automated recommendations.

## 5.1 Related Work

Trust is a complex concept. A vast body of literature on trust has grown in several areas of research but it is relatively confusing and sometimes contradictory, because the term is being used with a variety of meanings (McKnight and Chervany 2002). Golbeck (2005) defines trust as "trust in a person is a commitment to an action based on belief that the future actions of that person will lead to a good outcome". In reality the way a user decides about trust values in a real setting can depend on many factors, personal subjective tastes and previous experience. Some researchers have found that given a particular predefined domain and context, people's interest similarity is a strong predictor of interpersonal trust (Jensen et al. 2002; Golbeck and Hendler 2006). Ziegler and Golbeck (2007) have investigated the relationship between people's interest similarity and trust. Their empirical analysis on the correlation of interpersonal trust and interest similarity showed positive mutual interactions between interpersonal trust and interest similarity. That means, people who have similar interests tend to be more trusting towards each other. Under this assumption, we propose to construct user trustworthiness based on their interest similarity, generated from their tagging information.

Social tagging has drawn more and more attention recently. Many e-commerce websites or social networks are equipped with a tagging facility for users to classify or annotate their items. The current research on tags is mainly focused on how to build better collaborative tagging systems, personalize searches using tag information (Tso-Sutter et al. 2008) and recommending items (Sen et al. 2009) to users etc. However, tags are free-style vocabulary represented by text that users use to classify or label their items. As there is no restriction, boundary or pre-specified vocabulary regarding selecting words for tagging items, the tags used by users lack standardization and also contain a lot of ambiguity. Moreover, usually the tags are short containing only one or two words, which make it even harder to truly get the semantic meaning of the tags.

As tagging is a kind of implicit rating behavior (Sen et al. 2009) and the tags are pieces of textural information describing the content of items, mainly, the memory-based CF and content-based approaches can be used. Earlier work did not consider the tag quality problem (Liang 2011). Diederich et al. (2006) proposed an exploratory user-based CF approach using tag-based user profiles. The *tf-iuf* weighting approach similar to the *tf-idf* approach in text mining was used for each user's tags. The work of Tso-Sutter et al. (2008) extended the binary user-item matrix to binary user-item-tag matrix and used the Jaccard similarity measure approach to find neighbors. It was claimed that because of the tag quality problem, tag information failed to be very useful to improve the accuracy of memory-based CF approaches (Tso-Sutter et al. 2008).

More recently, the noise of tags or the quality and usefulness of tags aroused attention (Shepitsen et al. 2008; Sen et al. 2009; Liang 2011). Some content-based approaches that dealt with the noise of textural contents were proposed. In the work of Niwa et al. (2006) and Shepitsen et al. (2008), the clustering approaches were used to find the item and tag clusters based on the tag-based *tf-idf* content representations. The mapping of tags between users' tags and the representative tags of item clusters were used to make content-based web page recommendations. Approaches based on Latent Semantic Analysis, such as probabilistic Latent Semantic Index (PLSI) (Gemmis et al. 2008) and LDA (Siersdorfer et al. 2009), have been proposed to remove the noise of tags and build latent semantic topic models to recommend items to users.

Besides these memory-based CF approaches and content-filtering models, in the work of Sen et al. (2009), a special tag rating function was used to infer users' tag preferences. Along with the inferred tag preferences, the click streams and tag search history of each user was used to determine the user's preferences for items. The various kinds of extra information and special functions make the work of Sen et al. (2009) hard to compare with other studies on the subject and effectively restrict the applications of the work. More recently, Zhang et al. (2010) proposed to integrate the user-tag-item tripartite graph to rank items for the purpose of recommending unrated items to users. The user-tag-item graph was divided into user-tag and tag-item while the three-dimensional relationships reflecting the personal tagging relationships were ignored by the work of Zhang et al. (2010). Zhen et al. (2009) proposed to integrate tag information and explicit ratings to improve the accuracy of rating predictions of a model-based CF approach.

In this section, we propose to generate users' topic interests based on users' tagging behavior. The user's topic interests are then used to find a user's neighbors who share similar interests with the target user. To solve the quality problem of tags, we propose an approach to extract the semantic meaning of a tag based on the description of the items in that tag. For each item, we assume that there is a set of keywords or topics which describe the content of the item. This assumption is usually true in reality. For most products, normally there is a product description associated with the product. From the product description, by using text-mining techniques such as the *tf-idf* (Salton 1988) keywords extraction method, we can generate a set of keywords to represent the content of the item.

## 5.2 Similarity-Based Trust Construction

### 5.2.1 Notations

To describe the proposed similarity-based trust approach *SimTrust*, we define some concepts used in this section below.

- **Users:** $U = \{u_1, u_2, \ldots, u_{|U|}\}$ contains all users in an online community who have used tags to label and organise items.
- **Items (i.e. Products, Resources):** $P = \{p_1, p_2, \ldots, p_{|P|}\}$ contains all items tagged by users in $U$. Items could be any type of online information resources or products in an online community such as web pages, videos, music tracks, photos, academic papers, documents and books, etc.

    Without losing generality, we can assume that for each item, there is a piece of text which describes the content of the item, such as the content of a Web page, the title of a video or music track, the product description for a product, the document itself or the abstract for a document. Let $d_i$ denote the text associated with item $p_i$, and $d_i$ is a set of words appearing in the text which can be treated as a document, then $D = \{d_1, d_2, \ldots, d_{|P|}\}$ is a set of documents. From $D$, by using text-mining or data-mining methods, we can extract a set of frequent terms or keywords $W = \{w_1, \ldots, w_m\}$ such that $W \cap d_i, \neq \varphi$ for $i = 1, \ldots, $ |P|.

- **Tags:** $T = \{t_1, t_2, \ldots, t_{|T|}\}$ contains all tags used by users in $U$. A tag is a piece of textural information given by users to label or collect items.
- **Tagging relationship:** The three-dimensional tagging relationship is defined as $r : U \times T \times P \to \{0, 1\}$. If a user $u_i$ collected one item $p_k$ with tag $t_j$, then $r(u_i, t_j, p_k) = 1$ otherwise, $r(u_i, t_j, p_k) = 0$.

    As mentioned before, we believe the trustworthiness between users is useful for making recommendations. However, the trust information is not always available, and even if available, it may change over time. In this research, we propose to automatically construct the trustworthiness between users based on users' online information and online behaviour.

### 5.2.2 Proposed Approach

In a tag, a set of items are gathered together according to a user's viewpoint. We believe that there must be some correlation between the user's tag and the content of the items in that tag. Otherwise the user would not classify the items into that tag. Thus, using text-mining techniques, from the descriptions of the items in the tags, we can derive a set of keywords or topics to represent the semantic meaning of each tag.

For user $u_i \in U$, let $T_i = \{t_{i1}, \ldots, t_{il}\} \subseteq T$ be a set of tags that are used by $u_i$, $D_{t_{ij}}^{u_i}$ be a set of documents for tag $t_{ij} \in T_i$, similar to generating the keywords $W$ for $D$, by using text-mining techniques, from $D_{t_{ij}}^{u_i}$, we can extract a set of terms or keywords denoted as $W_{t_{ij}}^{u_i} = \{w_{ij}^1, \ldots, w_{ij}^n\}$, $W_{t_{ij}}^{u_i} \subseteq W$, and the corresponding normalised term frequency $F_{t_{ij}} = \{f_{ij}^1, \ldots, f_{ij}^n\}$ such that $\sum_{k=1}^n f_{ij}^k = 1$ to represent the semantic meaning of the tag $t_{ij}$. $W_{u_i} = \bigcup_{t_{ik} \in T_i} W_{t_{ik}}^{u_i}$ contains the terms or keywords for the tags used by user $u_i$. Let $S_{u_i}(t_{ij})$ denote the semantic representation of tag $t_{ij}$ for user $u_i$, $S_{u_i}(t_{ij})$ can be constructed as below:

$$S_{u_i}(t_{ij}) = \{(w_k, f_k)| \quad if \ w_k \in W_{t_{ij}}^{u_i} \wedge w_k = w_{ij}^l, \quad then f_k = f_{ij}^l,$$
$$otherwise f_k = 0, \ k = 1, \ldots, m\}$$

The semantic representation of a tag can be treated as an $m$-sized vector $<f_1, \ldots, f_m>$. By using the semantic representation, we can evaluate the similarity of two tags by calculating the similarity of the two corresponding vectors. In this book, the cosine similarity method is used to calculate the similarity of two tags. For two tags $t_p$ and $t_q$, their representations are $<f_p^1, \ldots, f_p^n>$ and $<f_q^1, \ldots, f_q^n>$, respectively, the similarity of the two tags is calculated using the following equation:

$$sim(t_p, t_q) = \frac{\sum_{i=1}^n (f_p^i * f_q^i)}{\sqrt{\sum_{i=1}^n (f_p^i)^2} * \sqrt{\sum_{i=1}^n (f_q^i)^2}} \tag{5.1}$$

The following is an example of calculating the semantic representations of tags. Let's assume that there are five users $u_1, u_2, \ldots, u_5$ who have used 15 items or products $p_1, p_2, \ldots, p_{15}$, tagged those products with 10 different tags $t_1, t_2, \ldots, t_{10}$. There are also 12 keywords $w_1, w_2, \ldots, w_{12}$ taken from f of the products. So, the scenario can be listed as follows:

User $U = \{u_1, u_2, u_3, u_4, u_5\}$
Item $P = \{p_1, p_2, p_3, p_4, p_5, p_6, p_7, p_8, p_9, p_{10}, p_{11}, p_{12}, p_{13}, p_{14}, p_{15}\}$
Tag $T = \{t_1, t_2, t_3, t_4, t_5, t_6, t_7, t_8, t_9, t_{10}\}$
Keyword $W = \{w_1, w_2, w_3, w_4, w_5, w_6, w_7, w_8, w_9, w_{10}, w_{11}, w_{12}\}$ where $m = 12$

From the taxonomy or description of each item, we can generate keywords for the items as shown in the Table 5.1:

From the user behavior of tagging; we may have the information about which user uses which items and tagged with which tags. The three-dimensional information $r : U \times T \times P$ looks like the Table 5.2:

In Table 5.2, for each user, the second column lists the tags used by the user and the third column lists the items collected under the corresponding tag by this user.

**Table 5.1** Item keywords list

| Item | Keywords |
|------|----------|
| $p_1$ | $w_1, w_2, w_3$ |
| $p_2$ | $w_1, w_2, w_4$ |
| $p_3$ | $w_5, w_6, w_7$ |
| $p_4$ | $w_5, w_6, w_8$ |
| $p_5$ | $w_9$ |
| $p_6$ | $w_9, w_{10}$ |
| $p_7$ | $w_9, w_{10}, w_{11}$ |
| $p_8$ | $w_9, w_{10}, w_{11}, w_{12}$ |
| $p_9$ | $w_1, w_4$ |
| $p_{10}$ | $w_2$ |
| $p_{11}$ | $w_{10}, w_{11}$ |
| $p_{12}$ | $w_{10}, w_{12}$ |
| $p_{13}$ | $w_{11}, w_{12}$ |
| $p_{14}$ | $w_{12}$ |
| $p_{15}$ | $w_{10}, w_{11}, w_{12}$ |

**Table 5.2** User tagging behavior

| User | Tag | Items |
|------|-----|-------|
| $u_1$ | $t_1$ | $p_1, p_2$ |
|  | $t_2$ | $p_3, p_{10}$ |
| $u_2$ | $t_3$ | $p_4, p_5$ |
|  | $t_6$ | $p_{11}, p_{12}$ |
| $u_3$ | $t_4$ | $p_1, p_4$ |
|  | $t_2$ | $p_5$ |
|  | $t_9$ | $p_1, p_2$ |
| $u_4$ | $t_5$ | $p_6, p_7, p_8, p_9$ |
|  | $t_8$ | $p_6$ |
|  | $t_9$ | $p_6$ |
| $u_5$ | $t_{10}$ | $p_{13}, p_{14}, p_{15}$ |

#### 5.2.2.1 Example of Tag Similarity Calculation

We show below how to calculate the tag representations by using tag $t_1$ for $u_1$ as an example. For user $u_1$, there are two items, $p_1$ and $p_2$, collected under tag $t_1$. So, $p_1$ and $p_2$ form a small transaction collection from which we can generate the term frequency (i.e. count ÷ size of the transaction collection) for the keywords of the two items. There are four keywords $w_1, w_2, w_3, w_4$ for the two items. The term frequency for $w_1, w_2, w_3, w_4$ would be 1, 1, 0.5, and 0.5, respectively. The normalized term frequency would be 0.33, 0.33, 0.17, and 0.17, respectively. Similarly, for tag $t_2$, the normalized term frequency for keywords $w_2, w_5, w_6, w_7$ would be 0.25, 0.25, 0.25, and 0.25, respectively. Thus, a vector of size 12 can be generated to represent the semantic meaning of each tag. For user $u_1$, the representation for $t_1$ and $t_2$ would be <1,1,0.5,0.5,0,0,0,0,0,0,0,0> and <0,0.25, 0,0,0.25,0.25,0.25,0,0,0,0,0> respectively.

**Table 5.3** Tag representations by normalized word frequencies

| User | Tag | Tag representation | | | | | | | | | | | |
|------|-----|-------|-------|-------|-------|-------|-------|-------|-------|-------|----------|----------|----------|
| | | $w_1$ | $w_2$ | $w_3$ | $w_4$ | $w_5$ | $w_6$ | $w_7$ | $w_8$ | $w_9$ | $w_{10}$ | $w_{11}$ | $w_{12}$ |
| $u_1$ | $t_1$ | 0.33 | 0.33 | 0.17 | 0.17 | 0 | 0 | 0 | 0 | 0 | 0 | 0 | 0 |
| | $t_2$ | 0 | 0.25 | 0 | 0 | 0.25 | 0.25 | 0.25 | 0 | 0 | 0 | 0 | 0 |
| $u_2$ | $t_3$ | 0 | 0 | 0 | 0 | 0.20 | 0.25 | 0 | 0.25 | 0.25 | 0 | 0 | 0 |
| | $t_6$ | 0 | 0 | 0 | 0 | 0 | 0 | 0 | 0 | 0 | 0.50 | 0.25 | 0.25 |
| $u_3$ | $t_4$ | 0.17 | 0.17 | 0.17 | 0 | 0.17 | 0.17 | 0 | 0.17 | 0 | 0 | 0 | 0 |
| | $t_2$ | 0 | 0 | 0 | 0 | 0 | 0 | 0 | 0 | 1 | 0 | 0 | 0 |
| | $t_9$ | 0.33 | 0.33 | 0.17 | 0.17 | 0 | 0 | 0 | 0 | 0 | 0 | 0 | 0 |
| $u_4$ | $t_5$ | 0.09 | 0 | 0 | 0.09 | 0 | 0 | 0 | 0 | 0.27 | 0.27 | 0.19 | 0.09 |
| | $t_8$ | 0 | 0 | 0 | 0 | 0 | 0 | 0 | 0 | 0.50 | 0.50 | 0 | 0 |
| | $t_9$ | 0 | 0 | 0 | 0 | 0 | 0 | 0 | 0 | 0.50 | 0.50 | 0 | 0 |
| $u_5$ | $t_{10}$ | 0 | 0 | 0 | 0 | 0 | 0 | 0 | 0 | 0 | 0.17 | 0.33 | 0.50 |

Following the same calculation procedure above, we can generate the tag representation for each tag used by the users in this example, as shown below in Table 5.3.

$\forall u_i, u_j \in U$, let $T_i = \{t_{i1}, \ldots, t_{il}\}, T_j = \{t_{j1}, \ldots, t_{jk}\} \subseteq T$ be the set of tags which were used by users $u_i$ and $u_j$, respectively. Corresponding to $T_i$ and $T_j$, $S_i = \{S_{u_i}(t_{i1}), \ldots, S_{u_i}(t_{il})\}$ and $S_j = \{S_{u_j}(t_{j1}), \ldots, S_{u_j}(t_{jk})\}$ are the tag semantic representations for the tags in $T_i$ and $T_j$, respectively. For tag $t_{ir}$ and tag $t_{is}$, let $sim(t_{ir}, t_{is})$ denote the similarity between $t_{ir}$ and $t_{is}$, if $sim(t_{ir}, t_{is})$ is larger than a pre-specified threshold, the two tags $t_{ir}$ and tag $t_{is}$ are considered similar.

Let $p(u_i \mid u_j)$ denote the likelihood that user $u_i$ is similar to user $u_j$ in terms of user $u_j$'s information interests. The following equation is defined to calculate how similar user $u_i$ is interested in keyword $w_k$ given that user $u_j$ is interested in the keyword $w_k$:

$$p_k(u_i|u_j) = \frac{n_{ij}^k}{n_j^k} \tag{5.2}$$

where, $n_j^k$ denotes the number of tags in $T_j$ that contain keyword $w_k$, $n_{ij}^k$ denotes the number of tags in $T_i$ that contain keyword $w_k$ and are similar to some tag in $T_j$ that contains keyword $w_k$ as well. $n_j^k$ and $n_{ij}^k$ are calculated by the following equations:

$$T_j^k = \{t_p | t_p \in T_j, w_k \in W_{t_p}\} \tag{5.3}$$

$$T_{ij}^k = \{t_p | t_p \in T_i, \exists t_q \in T_j, sim(t_p, t_q) > \sigma \wedge w_k \in W_{t_p} \wedge w_k \in W_{t_q}\} \tag{5.4}$$

where $\sigma$ is a pre-defined threshold to determine the similarity between tags.

$$n_j^k = |T_j^k| \tag{5.5}$$

$$n_{ij}^k = |T_{ij}^k| \tag{5.6}$$

After calculating $p_k(u_i|u_j)$ for every keyword, the average of the probability $p_k(u_i \mid u_j)$ is used to estimate the probability $p(u_i \mid u_j)$:

$$p(u_i|u_j) = \frac{\displaystyle\sum_{w_k \in W_{u_i,u_j}} p_k(u_i|u_j)}{|W_{u,u_j}|} \tag{5.7}$$

where $W_{u_i,u_j} = W_{u_i} \cup W_{u_j}$ is the set of all keywords contained in the tags used by $u_i$ or $u_j$. In this book, we use the conditional probability $p(u_i \mid u_j)$ to measure the trust from user $u_j$ to user $u_i$. Given $u_j$, the higher the $p(u_i \mid u_j)$, the more user $u_j$ trusts $u_i$ since user $u_i$ has similar interests to $u_j$. Because $p(u_i \mid u_j)$ represents averagely how much a user $u_i$ is interested in some topics (i.e. the keywords in the context of this book) given that another user $u_j$ is interested in these topics, high $p(u_i \mid u_j)$ indicates that the interest of user $u_i$ is similar to that of user $u_j$ in terms of these topics. On the other hand, low $p(u_i \mid u_j)$ indicates that user $u_i$ does not share the same interests with user $u_j$ in terms of these topics. Based on the findings discussed in Chap. 4 that the correlation between trust and interest similarity is positive, in this book, we use $p(u_i \mid u_j)$ to represent the trust from $u_i$ to $u_j$. The algorithms to calculate the trust between two users are given below.

---

**Algorithm 5.1 *SimTrust* ($u_a$, $u_b$)**

---

**Input:** $u_a, u_b \in U$ are two users

**Output:** $Trust(u_a, u_b)$, the trust from $u_a$ to $u_b$

**Begin**

1:   $Trust(u_a, u_b) := 0$ //initialisation

2:   **for** each keyword $w_k \in W_{u_a} \cup W_{u_b}$ {

3:        $p_k(u_a \mid u_b) := Trust\_Keyword(u_a, u_b, w_k)$//call algorithm 5.2

4:        $Trust(u_a, u_b) := Trust(u_a, u_b) + p_k(u_a \mid u_b)$

     }//**end for**

5:   $Trust(u_a, u_b) := \dfrac{Trust(u_a, u_b)}{|W_{u_a} \cup W_{u_b}|}$

6:   **Return** $Trust(u_a, u_b)$

**End**

---

---

**Algorithm 5.2 *Trust_Keyword*** $(u_a, u_b, w_k)$

---

**Input:** $u_a, u_b \in U$ are two users

$w_k \in W_{u_a} \cup W_{u_b}$ is a keyword contained in the tags used by $u_a$ or $u_b$

**Output:** *Trust_k*$(u_a, u_b)$, the trust from $u_a$ to $u_b$ in terms of keyword $w_k$

**Begin**

1:   $N_b := 0$, $N_{a,b} := 0$ //initialisation

2:   **for** each tag $t_p \in T_{u_b}$ {

3:       **if** $w_k \in W_{t_p}^{u_b}$, **then** $N_b := N_b + 1$   }//**end for**

4:   **for** each tag $t_p \in T_{u_a}$ {

//the following two variables are used to represent

//the current most similar tag and the similarity value

*mostSimilarValue*:=0

*mostSimilarTag*:=*null*

5:       **for** each tag $t_q \in T_{u_b}$ {

6:           **if** $w_k \in W_{t_p}^{u_a} \wedge w_k \in W_{t_q}^{u_b} \wedge (sim(t_p, t_q) > \sigma)$ **then** {

**if** $sim(t_p, t_q) > mostSimilarValue$ **then** {

}//**end**                                    }//**end for**

**if** $mostSimilarValue \neq 0$ **then** {

$N_{a,b} := N_{a,b} + 1$

$T_{u_b} := T_{u_b} \setminus mostSimilarTag$ //remove *mostSimilarTag* from $T_{u_{bf}}$

} //**end if**

}//**end for**

7: **if** $N_b = 0$ **then** *Trust_k*$(u_a, u_b)$ :=0

**else** *Trust_k*$(u_a, u_b) := \dfrac{N_{a,b}}{N_b}$

8: **Return** *Trust_k*$(u_a, u_b)$

**End**

**Table 5.4**  Tags and keywords used by user $u_1$ and $u_2$

| User | Tags | Keywords |
|------|------|----------|
| $u_1$ | $t_1, t_2$ | $w_1, w_2, w_3, w_4, \mathbf{w_5}, \mathbf{w_6}, w_7$ |
| $u_2$ | $t_3, t_6$ | $\mathbf{w_5}, \mathbf{w_6}, w_8, w_9, w_{10}, w_{11}, w_{12}$ |

#### 5.2.2.2  Computation Complexity

The computation complexity for Algorithm 5.2 is $O(l_1 \times l_2)$ where $l_1$ is the number of tags used by user $u_a$ and $l_2$ is the number of tags used by user $u_b$. Let $l = |W_{u_a} \cup W_{u_b}|$, the computation complexity for Algorithm 5.1 is $O(l \times l_1 \times l_2)$. Usually, $l_1 \ll |T|$, $l_2 \ll |T|$, and $l \ll |W|$.

#### 5.2.2.3  Example of User Similarity Calculation

The following Table 5.4 lists the tags and the keywords which appear in the corresponding tags of users' $u_1$ and $u_2$.

The set of keywords used by either $u_1$ or $u_2$ is $w_1 \cup w_2 = \{w_1, w_2, w_3, w_4, w_5, w_6, w_7, w_8, w_9, w_{10}, w_{11}, w_{12}\}$. It can be easily found that only $w_5$ and $w_6$ are included in both users' tags. Therefore, for the other keywords, $p_{w_k}(u_2|u_1)=0$, where $k = 1, 2, 3, 4, 7, 8, 9, 10, 11, 12$.

Below we will give the detail to calculate $p_{w_5}(u_2|u_1)$ and $p_{w_6}(u_2|u_1)$ using algorithm 5.2.

- $p_{w_5}(u_2|u_1)$

For user $u_1$, the number of tags which contain $w_5$ (which is $t_2$) is $n_1^{w_5} = 1$
For user $u_2$, only tag $t_3$ contains $w_5$.
So, the similarity between $u_1$'s $t_2$ and $u_2$'s $t_3$ can be calculated using the Eq. 5.1.

$$sim(t_2, t_3) = \frac{\sum_{i=1}^{12}(f_2^i * f_3^i)}{\sqrt{\sum_{i=1}^{12}(f_2^i)^2} * \sqrt{\sum_{i=1}^{n}(f_3^i)^2}} = 0.5$$

If we consider the threshold value for this example is 0.5 (in our actual experiment the threshold value is set as 0.8), then $Sim(t_2, t_3) = 0.5$ indicates that $u_1$'s $t_2$ and $u_2$'s $t_3$ is similar, which results in $n_{1,2}^{w_5} = 1$

$$P_{w_5}(u_2|u_1) = 1/1 = 1$$

- $p_{w_6}(u_2|u_1)$

For user $u_1$, the number of tags which contain $w_6$ (which is $t_2$) is $n_1^{w_6} = 1$

For user $u_2$, only tag $t_3$ contains $w_6$ and the similarity between $u_1$'s $t_2$ and $u_2$'s $t_3$ is also

$$Sim(t_2, t_3) = 0.5$$

Similarly, for the threshold 0.3, $Sim(t_2, t_3) = 0.5$ indicates that $u_1$'s $t_2$ and $u_2$'s $t_3$ is similar, which results $n_{1,2}^{w_6} = 1$

$$P_{w6}(u_2|u_1) = 1/1 = 1$$

Therefore

$$p(u_2|u_1) = (0+0+0+0+1+1+0+0+0+0+0+0)/12 = 0.166$$

This means that the similarity between $u_1$ and $u_2$ based on the tag similarity is 0.166.

## 5.3  Chapter Summary

This chapter has presented a new algorithm for generating trust networks based on user tagging information to make recommendations in the online environment. The algorithm has described in this chapter with numerical example derivation for both, the tag similarity and user similarity calculation. The experiment results which are presented in detail in the next chapter showed that this tag-based similarity approach performs better while making recommendations than the traditional collaborative filtering based approach. This proposed technique will be very helpful to deal with data sparsity problem; even when explicit trust rating data is not available. The finding will contribute in the area of recommender system by improving the overall quality of automated recommendation making.

# Chapter 6
# Experiments and Evaluation

**Abstract** This chapter presents how the tag-similarity based trust-network recommendation and the proposed directed serial parallel graph trust transitivity approaches are evaluated. Firstly, the experimental method and the analysis of the dataset is presented, then the baseline model of similarity calculation and recommendation techniques, followed by a description of the evaluation metrics, finally the experiment results are presented with detailed analysis and discussion.

**Keywords** Recommender systems · Trust · Inference · User tag · Web 2.0 · Social networks

This chapter presents how the tag-similarity based trust-network recommendation and the proposed directed serial parallel graph trust transitivity approaches are evaluated. Firstly, the experimental method and the analysis of the dataset is presented, then the baseline model of similarity calculation and recommendation techniques, followed by a description of the evaluation metrics, finally the experiment results are presented with detailed analysis and discussion.


## 6.1 Experimental Method

For evaluating recommender systems, many different techniques and systems have been proposed and developed. Usually, a recommender system can be evaluated by its recommendation quality (how many correct recommendations it can generate) or computation efficiency (how quickly it can generate recommendations). In the first approach, a recommender is evaluated based on whether its recommendations can satisfy users' information needs, which means, if the recommender's recommendation is good, then it would make most of its users happy and satisfied (Herlocker et al. 2004). In the second approach, the computation efficiency evaluation aims to ensure a recommender has the ability to handle large numbers of recommendation requests in real time (Rashid et al. 2006; Sarwar et al.

T. Bhuiyan, *Trust for Intelligent Recommendation*,
SpringerBriefs in Electrical and Computer Engineering,
DOI: 10.1007/978-1-4614-6895-0_6, © The Author(s) 2013

2002). A common approach to evaluate a recommender's computation efficiency is to measure the amount of time it requires to generate a single recommendation. Most of the studies in this field consider a recommender's recommendation quality over its computation efficiency. It makes sense because the efficiency bottleneck can be solved by many non-algorithmic approaches such as employing higher performance hardware by using faster processors, additional primary memory etc., (Karypis 2001; Sarwar et al. 2000). But, recommendation quality can only be improved via using better algorithms. In this research work, evaluation will focus on recommendation quality based on the number of correct recommendations the system can generate rather than the computation efficiency of its required time to generate a recommendation.

The objectives of the experiments are to verify the effectiveness of the proposed approaches. To achieve this, the experiments are conducted based on the following hypotheses.

- **Hypothesis 1**: The recommendation quality produced by using the proposed tag-based similarity approach *SimTrust* is better than that of using the traditional Jaccard similarity approach *JacSim*.
- **Hypothesis 2**: The recommendation quality, is improved when the range of the neighbourhood is expanded using the trust inference algorithm, compared to the traditional collaborative filtering technique using the Jaccard similarity or using the proposed tag-based similarity approach.
- **Hypothesis 3**: The proposed trust inference algorithm *DSPG* outperforms other popular trust inference algorithms.

The similarity calculation of Jaccard is totally based on the previous common behavior. Whereas our proposed tag based similarity approach use the users' interest similarity or tests, which should produce better recommendation. This idea leads to set up hypothesis 1. With sufficient neighbor information, the collaborative filtering works well. As the lack of sufficient neighbor information is common in real world, increasing the boundary or range of neighborhood will increase the possibility of improving recommendation quality. Based on this concept, the hypothesis 2 has set. Our proposed *DSPG* algorithm used an innovative way to propagate trust value within a trust network to expand the neighborhood which has reason to believe that will perform better than the other common inference algorithms. Because of this, hypothesis 3 has set.

We have used the traditional collaborative filtering algorithm to make recommendations. The traditional collaborative filtering algorithm has two steps. First, it finds the similar neighbors based on the overlap of previous ratings and in the second step, it calculates to predict an item to recommend to a target user. For all of the experimental data, we have used the same method for the second part of the algorithm. But, we have used our proposed trust network to find the neighbors and make recommendations, then compared those recommendation results with the traditional collaborative filtering method using Jaccard's coefficient to find the neighbors. To verify each of the hypotheses, we have used the Amazon book data collected from www.amazon.com. The effectiveness of the evaluation is the main focus of this chapter which is discussed in detail in the following sections.

## 6.2 Experimental Dataset

For our experiment, we have used a real-world book dataset downloaded from the popular web site www.amazon.com. User tag data is also obtained from the Amazon site. The dataset was downloaded in April 2008. The book data has some significant differences to other data about movies, games or videos. Every published book has a unique ISBN, which makes it easy to ensure interoperability and gather supplementary information from various other sources, e.g., taxonomy or category descriptors from Amazon for any given ISBN, and thus it is most suitable for evaluating our proposed method. The dataset consists of 3,872 unique users, 29,069 books, 84,206 tags and a total of 14,658 keywords. We have also extracted the book taxonomy descriptors of each book from the same site (Ziegler and Lausen 2004; Weng et al. 2008). Amazon's book taxonomy is a tree structure. It contains 9,919 unique topics and the average number of descriptors per book is around 3.15. Each book in the Amazon dataset has several taxonomic descriptors, each of which is a set of categories in the book taxonomy. An example of a small fragment of the book taxonomy extracted from www.amazon.com is shown in Fig. 6.1.

In this experiment, we have extracted keywords for each tag from the descriptors of the books in the tag. For example, user "A11SWG9T60IQH8" has used 2 items (i.e., books) whose ids are "1887166386" and "1887166440" and tagged these items with the tag "girl stuff". From the Amazon book taxonomy, we can find the descriptors for the book "1887166386" include "Subjects > Entertainment > Humor > General;    Subjects > Literature    and    Fiction > World

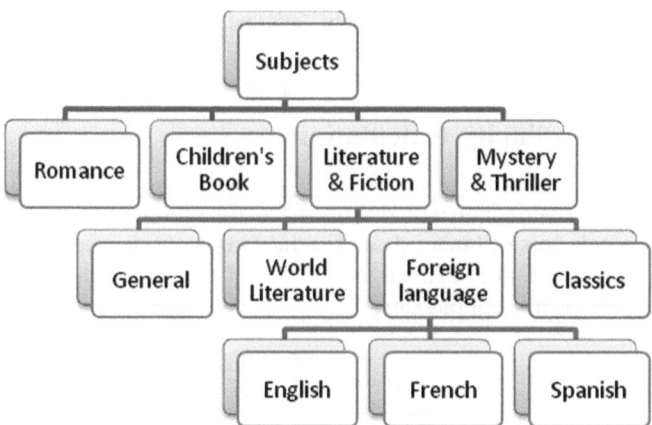

**Fig. 6.1** An example fragment of item taxonomy

Literature > British > 20th Century; Subjects > Reference > Quotations;" and
the descriptors for the book "1887166440" include "Subjects > Entertain-
ment > Humor > General; Subjects > Entertainment > Humor > Love, Sex and
Marriage;   Subjects > Entertainment > Humor > Parodies;   Subjects > Health,
Mind and Body > Relationships > Interpersonal Relations; Subjects > Health,
Mind and Body > Relationships > Love and Romance; Subjects > Home and
Garden > Crafts   and   Hobbies > Sewing;   Subjects > Nonfiction > Social
Sciences > Sociology > Marriage and Family".

From these descriptors, we may generate the keyword set for the tag "girl stuff"
for the user "A11SWG9T60IQH8" as "Entertainment, Humor, General, Literature
and Fiction, World Literature, British, 20th Century, Reference, Quotations Love,
Sex and Marriage, Parodies, Health, Mind and Body, Relationships, Interpersonal
Relations, Love and Romance, Home and Garden, Crafts and Hobbies, Sewing,
Nonfiction, Social Sciences, Sociology, Marriage and Family".

## 6.3 Parameterization

To calculate the neighbor of a given user in the network, we use the *SimTrust*
program. There are two thresholds $\sigma_1$ and $\sigma_2$ used to perform the neighbor formation experiment. $\sigma_1$ is the threshold to find the similar tags where the name or
textual information of two tags may be different but the contents are close. For our
experiment we set the value for $\sigma_1$ to 0.8. The other threshold $\sigma_2$ is threshold to
determine similar users. The value for $\sigma_2$ is set to 0.4 in this experiment. We
performed some pre-processing for our experiment to select the suitable threshold
$\sigma$ value for this particular set of experimental data. The value should be chosen
individually for different datasets depending on the objectives. *MoleTrust* requires
a threshold to control the number of hops to search from the source to the target
node to find the possible trust paths. In this experiment, we use the number of hops
$\vartheta = 4$. For $t$ test evaluation, we have used *"paired"* and *"one-tailed distribution"*.

## 6.4 Jaccard Similarity Calculations

Jaccard similarity coefficient is a statistic used for comparing the similarity and
diversity of sample sets of binary data. The Jaccard coefficient measures similarity
between sample sets, and is defined as the size of the intersection divided by the
size of the union of the sample sets:

$$J(A, B) = \frac{|A \cap B|}{|A \cup B|} \tag{6.1}$$

To calculate the Jaccard similarity coefficient, we use the following formula:

Let $p$ be the number of items that are rated by both users $u_i$ and $u_j$; $q$ be the number of items that are rated by $u_i$ but not by $u_j$; $r$ be the number of items that are not rated by $u_i$ but rated by $u_j$, the Simple Matching version of the Jaccard coefficient for measuring the similarity between users $u_i$ and $u_j$ can be expressed as:

$$S_{ij} = \frac{p}{p + q + r} \qquad (6.2)$$

where;

$p =$ number of variables that are positive for both objects
$q =$ number of variables that are positive for the $i$th objects and negative for the $j$th object
$r =$ number of variables that are negative for the $i$th objects and positive for the $j$th object
$S =$ number of variables that are negative for both objects.

## 6.5   Traditional Approach of Collaborative Filtering

Collaborative filtering recommender systems try to predict the utility of items for a particular user based on the items previously rated by other users (Adomavicius and Tuzhilin 2005). More formally, the utility, denoted as $u(c,s)$ for item $s$ and for user $c$, is estimated based on the utilities $u(c_j,s)$ assigned to item $s$ by those users $c_j \in C$ who are "similar" to user $c$. The task of collaborative filtering is to predict the value of the unknown rating $r_{c,s}$ for user $c$ and item $s$ which is usually computed as an aggregate of the ratings of some other users (i.e., the neighbor users) for the same item $s$:

$$r_{c,s} = aggr_{c' \in C} r_{c's} \qquad (6.3)$$

where $C$ denotes the set of $N$ neighbor users that are the most similar to user $c$ and who have rated item s ($N$ can range anywhere from 1 to the number of all users). We have used the following aggregation function:

$$r_{c,s} = k \sum_{c' \in C} sim(C, C') \times r_{c',s} \qquad (6.4)$$

where multiplier $k$ serves as a normalizing factor and is selected as

$$k = \frac{1}{\sum_{c' \in C} sim(c, c')} \qquad (6.5)$$

Users' implicit ratings are usually binary values, i.e., value 1 represents that an item is rated and value 0 means that the item is not rated by the user. In the case of binary ratings, the Jaccard coefficient is usually used to measure the similarity between two users.

## 6.6  Evaluation Methods

The "Precision, Recall and *F1*" metrics are used to evaluate the recommendation performance. This evaluation method was initially suggested by Cleverdon as an evaluation metric for information retrieval systems (Cleverdon et al. 1966). Due to the simplicity and the popular use of these metrics, they have been also adopted for recommender system evaluations (Basu et al. 1998; Billsus and Pazzani 1999; Sarwar et al. 2000). The top-*N* items are recommended to the users. For comparison, we have used $N = 5$, 10, 15, 20, 25 and 30. The training and testing datasets both contain users, books and tag information. Precision and Recall for an item list recommended to user $u_i$ is computed based on the following equations:

$$\text{Precision} = \frac{|T_i \cap P_i|}{|P_i|} \tag{6.6}$$

$$\text{Recall} = \frac{|T_i \cap P_i|}{|T_i|} \tag{6.7}$$

where $T_i$ is the set of all items preferred by user $u_i$ and $P_i$ is the set of all recommended items generated by the recommender system. Precision is defined as the ratio between the number of recommended relevant items and the number of recommended items. It represents the probability that a recommended item is relevant. It can be seen as a measure of exactness or conformity. On the other hand, Recall is defined as the ratio between the number of recommended relevant items and the total number of relevant items that actually exist. It represents the probability that a relevant item will be recommended. It can be seen as a measure of completeness. Based on the Eqs. 6.6 and 6.7, it can be observed that the values of precision and recall are sensitive to the size of the recommended item list. Since, precision and recall are inversely correlated and are dependent on the size of the recommended item list, they must be considered together to completely evaluate the performance of a recommender (Herlocker et al. 2004). *F1* Metric is one of the most popular techniques for combining precision and recall together in recommender system evaluation. *F1*, which can be computed by the formula 6.8, is used for our evaluation. It provides a general overview of the overall performance.

$$F1 = \frac{2 \times \text{Precision} \times \text{Recall}}{\text{Precision} + \text{Recall}} \tag{6.8}$$

**Fig. 6.2** Fivefold cross-validation techniques

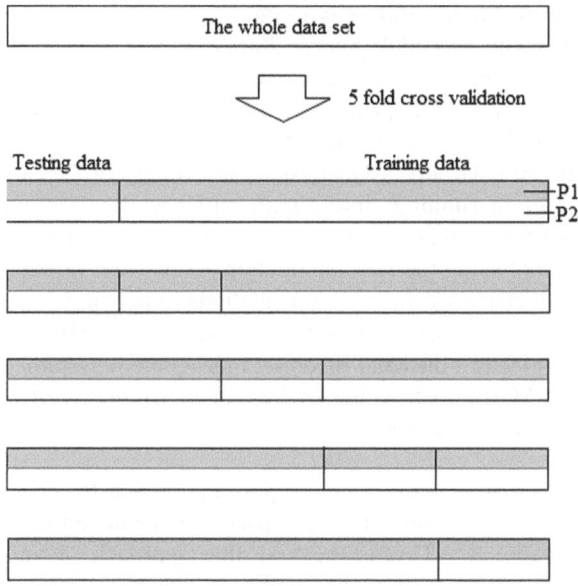

Cross-validation or rotation estimation is a statistical method of evaluating and comparing learning algorithms by dividing data into two segments: one used to learn or train a model and the other used to validate the model. In typical cross-validation, the training and validation sets must cross over in successive rounds such that each data point has a chance of being validated against. For our experiment; fivefold cross-validations are used to evaluate the effectiveness of the proposed approach. The whole dataset has randomly been portioned into five sub datasets. Among the five datasets, one single sub dataset (which in our case is 20 % of the total users) was retained as the validation data for testing and the remaining four sub datasets (the remaining 80 % of the users) were used as training data. For each test user, randomly selected, 20 % of the items associated with the user were hidden as the test item set while 80 % of the user's items were used as that user's training item set. Figure 6.2 visualizes the basic mechanism of fivefold cross-validation which was used for the experiment.

The average precision, recall and *F1* measure of the whole test users of one portioned validation sub dataset has been recorded as one run of the results. The average precision and recall values of the five runs have been used to measure the accuracy performance of the recommendations. For the computation efficiency evaluation, the average time required by the inference algorithms for a single user has also been compared and presented in Fig. 6.12.

To evaluate the effectiveness of the proposed approaches, this book implemented the proposed trust-based similarity model *SimTrust* using user's tag information and the proposed trust inference algorithm *DSPG* together with other related base line models for similarity calculation and trust inferences. The experiment has been implemented in Microsoft.NET (dot net) platform using the

Visual Studio 2008 environment with language C#. For backend database management, MS SQL Server 2008 has been used. The entire experiment has been
conducted on a Personal Computer with the configuration of a Pentium IV 3.0 GH
CPU with 2 GB primary memory running XP operating system. The experiment
could have been conducted on a High Performance Computer (HPC) provided by
the HPC Group of Queensland University of Technology which could reduce the
program running time significantly as the HPC computers are equipped with
4–6 GB primary memory and better performance CPU's. There were two reasons
for not using the HPC computers. Firstly, our research focus was on improving the
correctness of the recommendations rather improving the speed of the processing
time. Secondly, at this point in time QUT HPC does not allow a SQL Server
DBMS as a backend database management system. We chose C# as the language
and SQL Server as the DBMS for its simplicity of implementation. The proposed
approaches include:

| | |
|---|---|
| *SimTrust* | This is the proposed approach for generating a similarity-based trust network using users' personalized tagging information |
| *JacSim* | This is the base line approach of finding similarity between users using the previous common item ratings popularly used for collaborative filtering based recommender systems |
| *TidalTrust* | This is the base line model used to compare the performance of trust inferences. This method is the most popular method for trust inference at this time and is widely used and referred to in the literature (Golbeck 2005; Massa and Avesani 2007; Ziegler and Golbeck 2007) |
| *MoleTrust* | This is another baseline model for trust inference. This model differs from *TidalTrust* in terms of limiting the number of hops while searching the trust paths in the network to a preset threshold value (Massa et al. 2005) |
| *DSPG* | This is the proposed trust inference algorithm using subjective logic |

For each test user, this book compared the recommendation results produced by
the proposed approaches with the two most popular inference algorithms as the
baseline model.

## 6.7  Experiment Results and Discussion: Hypothesis 1

To test our first hypothesis, we compared our proposed tag-based similarity trust
*SimTrust* with the traditional Jaccard coefficient based similarity. The result
showed in the precision, recall and *F1*-Measure metrics in Tables 6.1, 6.2 and 6.3
and the corresponding figures in Figs. 6.3, 6.4 and 6.5 clearly indicate that our
proposed *SimTrust* approach is much more effective than the Jaccard similarity

**Table 6.1** Average precision values for *Approach A* using proposed *SimTrust* algorithm and *Jaccard* algorithm

| Algorithm | Top 5 | Top 10 | Top 15 | Top 20 | Top 25 | Top 30 |
|-----------|-------|--------|--------|--------|--------|--------|
| *SimTrust* | 0.032 | 0.027 | 0.023 | 0.020 | 0.016 | 0.015 |
| *Jaccard* | 0.008 | 0.006 | 0.004 | 0.003 | 0.003 | 0.002 |

**Table 6.2** Average recall values for *Approach A* using proposed *SimTrust* algorithm and *Jaccard* algorithm

| Algorithm | Top 5 | Top 10 | Top 15 | Top 20 | Top 25 | Top 30 |
|-----------|-------|--------|--------|--------|--------|--------|
| *SimTrust* | 0.014 | 0.022 | 0.025 | 0.029 | 0.030 | 0.032 |
| *Jaccard* | 0.006 | 0.009 | 0.009 | 0.009 | 0.009 | 0.009 |

**Table 6.3** Average *F1*-Measure values for *Approach A* using proposed *SimTrust* algorithm and *Jaccard* algorithm

| Algorithm | Top 5 | Top 10 | Top 15 | Top 20 | Top 25 | Top 30 |
|-----------|-------|--------|--------|--------|--------|--------|
| *SimTrust* | 0.019 | 0.025 | 0.024 | 0.023 | 0.021 | 0.020 |
| *Jaccard* | 0.007 | 0.007 | 0.006 | 0.005 | 0.004 | 0.003 |

approach in terms of correct recommendation making. The following tables and the corresponding figures show the average precision, recall and *F1*-Measure values for the experiment conducted. For the evaluation of this experiment, we have used two different approaches and the results from both support our hypothesis 1 as being correct.

***Approach A***: Compare the recommended result with the actual item used by the user on the basis of exact hit or match.

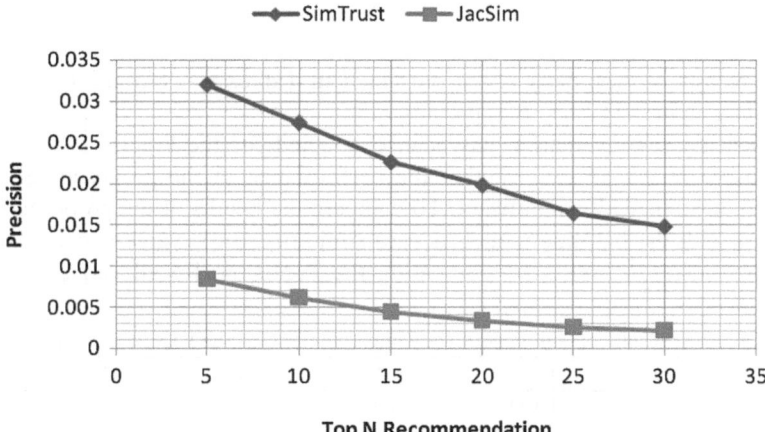

**Fig. 6.3** Evaluation for *Hypothesis 1 Approach A* with precision metric

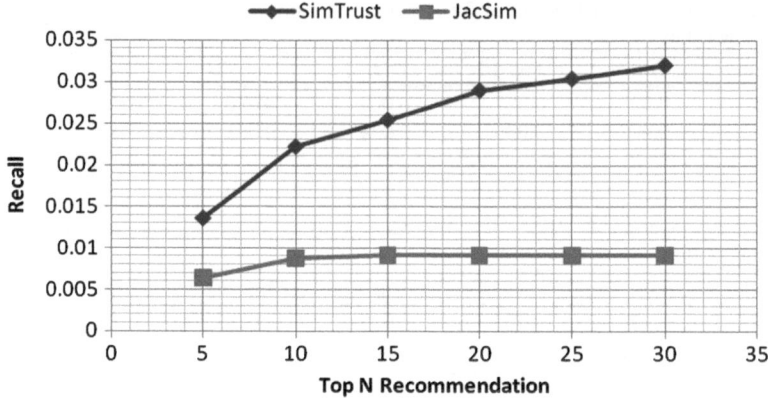

**Fig. 6.4** Evaluation for *Hypothesis 1 Approach A* with recall metric

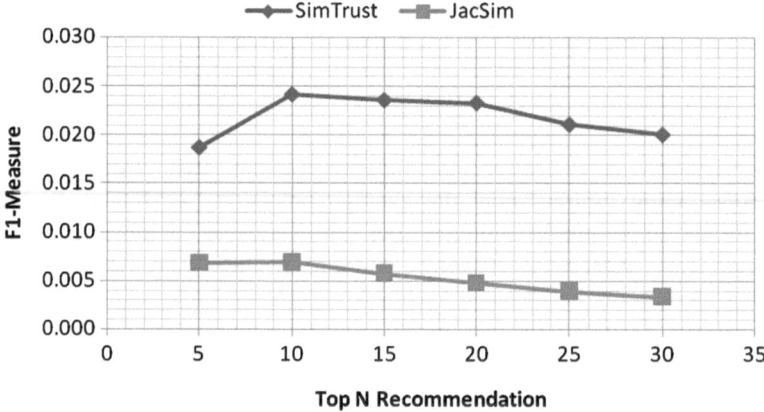

**Fig. 6.5** Evaluation for *Hypothesis 1 Approach A* with *F1*-Measure metric

The results of Approach A are presented in Tables 6.1, 6.2 and 6.3 and corresponding figures in Figs. 6.3, 6.4 and 6.5.

Student's $t$ test evaluation has been conducted for the Precision, Recall and *F1*-Measure results to test whether the performance of the proposed *SimTrust* approach is statistically significant. To compare the results of the proposed model *SimTrust* with the baseline model Jaccard similarity, the $p$ value is calculated for both approaches A and B by comparing the Precision, Recall and *F1*-Measure

**Table 6.4** *T*-test results of *Approach A* for comparing *SimTrust* and *Jaccard* algorithm

|            | Top 5      | Top 10     | Top 15     | Top 20     | Top 25     | Top 30     |
|------------|------------|------------|------------|------------|------------|------------|
| Precision  | 2.180E–03  | 1.192E–03  | 1.252E–03  | 1.244E–03  | 1.143E–03  | 1.045E–03  |
| Recall     | 1.043E–03  | 2.094E–03  | 2.698E–03  | 1.243E–03  | 1.130E–03  | 8.503E–04  |
| *F1*-Measure | 1.693E–03 | 8.989E–04  | 1.190E–03  | 1.346E–03  | 1.230E–03  | 9.288E–04  |

**Table 6.5** Average precision values for *Approach B* using proposed *SimTrust* algorithm and *Jaccard* algorithm

| Algorithm | Top 5 | Top 10 | Top 15 | Top 20 | Top 25 | Top 30 |
|-----------|-------|--------|--------|--------|--------|--------|
| *SimTrust* | 0.065 | 0.056 | 0.052 | 0.048 | 0.043 | 0.039 |
| *Jaccard* | 0.009 | 0.007 | 0.005 | 0.004 | 0.003 | 0.003 |

**Table 6.6** Average recall values for *Approach B* using proposed *SimTrust* algorithm and *Jaccard* algorithm

| Algorithm | Top 5 | Top 10 | Top 15 | Top 20 | Top 25 | Top 30 |
|-----------|-------|--------|--------|--------|--------|--------|
| *SimTrust* | 0.031 | 0.047 | 0.061 | 0.072 | 0.078 | 0.084 |
| *Jaccard* | 0.007 | 0.010 | 0.011 | 0.011 | 0.011 | 0.011 |

**Table 6.7** Average *F1*-Measure values for *approach B* using proposed *SimTrust* algorithm and *Jaccard* algorithm

| Algorithm | Top 5 | Top 10 | Top 15 | Top 20 | Top 25 | Top 30 |
|-----------|-------|--------|--------|--------|--------|--------|
| *SimTrust* | 0.042 | 0.051 | 0.056 | 0.058 | 0.055 | 0.053 |
| *Jaccard* | 0.007 | 0.008 | 0.007 | 0.006 | 0.005 | 0.005 |

values. If the $p$ value is less than 7.000E–02, the performance of the proposed approach has considered significantly improved. Table 6.4 shows the $t$ test evaluation results for proposed *SimTrust Approach A* and Table 6.8 shows for *Approach B*, compare to baseline model *Jaccard* similarity. The results show all the $p$ values of both approaches compared to Jaccard similarity model are less than 7.000E–02 for Precision, Recall and *F1*-Measure, which means both the proposed approaches are significantly outperform the baseline model.

***Approach B***: Instead of using the exact match or hit, first we calculate the target user's interested topics on the item and compare them with the recommender system generated items descriptors or topics. The overlaps of recommended and actual item topics are considered for the evaluation calculation.

The result of *Approach B* is presented in Tables 6.5, 6.6 and 6.7 and corresponding figures in Figs. 6.6, 6.7 and 6.8. Approach B performs much better than *Approach A* as the recommendation values are much higher. However, in both approaches, our proposed tag-based interest similarity trust approach outperforms the traditional Jaccard coefficient based similarity approach.

Table 6.8 shows the $t$ test results for *Approach B*, compare to baseline model *Jaccard* similarity. All the $p$ values of the above table are less than 7.000E–02 for Precision, Recall and *F1*-Measure, which means the proposed approach significantly outperforms the baseline model for *Approach B* as well.

It can be seen from the results shown in Tables 6.1, 6.2, 6.3, 6.5, 6.6 and 6.7 that the recommendation quality of the proposed *SimTrust* is much better than the *Jaccard* similarity approach. The $t$ test results for comparing *SimTrust* and *Jaccard* also confirms the significance of the outcomes. However, as *SimTrust* needs to use

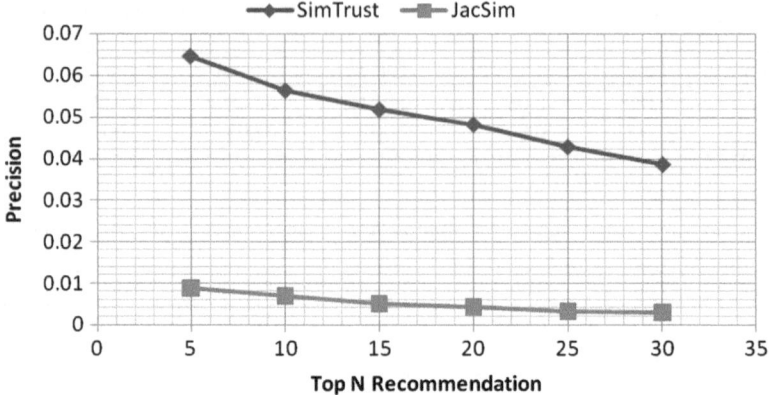

**Fig. 6.6** Evaluation for *Hypothesis 1 Approach B* with precision metric

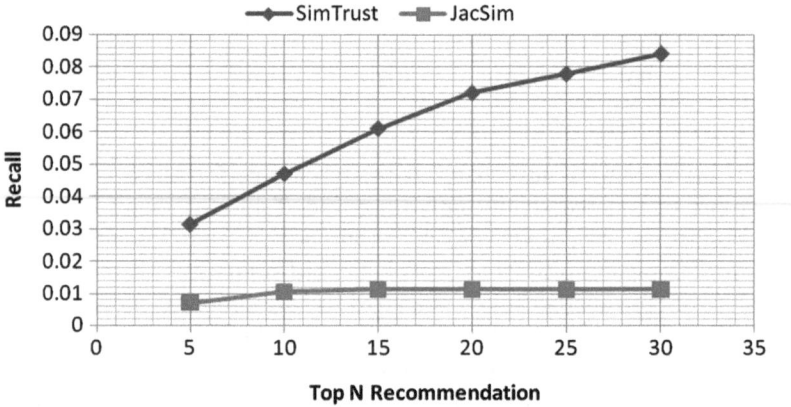

**Fig. 6.7** Evaluation for *Hypothesis 1 Approach B* with recall metric

the expensive tag-based similarity computation for all users within their neighborhood including the target user, it requires more time than the simple common item comparison of the *Jaccard* similarity calculation approach. This similarity calculation can be conducted offline which will have no effect on the actual time taken for making recommendations. The similarity between the users can be preprocessed before the online recommendation generation.

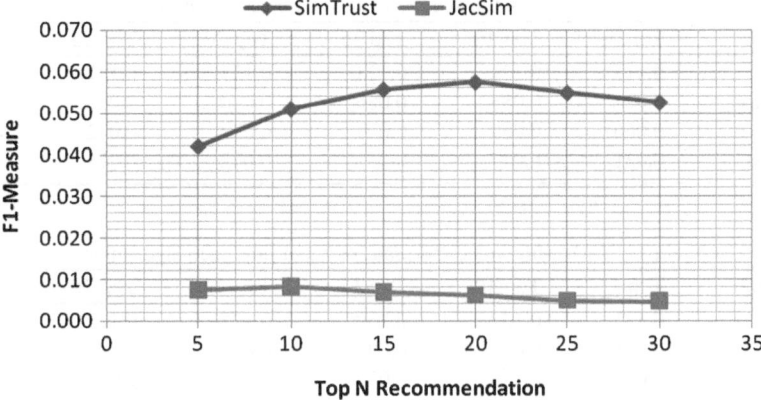

**Fig. 6.8**  Evaluation for *Hypothesis 1 Approach B* with *F1*-Measure metric

**Table 6.8**  *T*-test results of *Approach B* for comparing *SimTrust* and *Jaccard* algorithm

|            | Top 5     | Top 10    | Top 15    | Top 20    | Top 25    | Top 30    |
|------------|-----------|-----------|-----------|-----------|-----------|-----------|
| Precision  | 7.083E–05 | 5.315E–05 | 7.694E–05 | 3.866E–05 | 4.429E–05 | 4.609E–05 |
| Recall     | 2.378E–04 | 8.246E–05 | 7.679E–06 | 1.096E–05 | 8.869E–06 | 9.429E–06 |
| *F1*-Measure | 1.492E–04 | 4.528E–05 | 2.650E–05 | 1.970E–05 | 1.504E–05 | 1.786E–05 |

## 6.8  Experiment Results and Discussion: Hypothesis 2 and 3

For this experiment, we use the following four different techniques to calculate neighbor users for a given target user, then use the same traditional collaborative filtering technique to make recommendations.

- The proposed tag-based Similarity Trust method *SimTrust* without any propagation.
- Expand the neighbour range by propagating trust using Golbeck's *TidalTrust* algorithm.
- Expand the neighbour range by propagating trust using Massa's *MoleTrust* algorithm.
- Expand the *SimTrust* neighbours by using our proposed *DSPG* inference algorithm.

**Table 6.9**  Average precision values for *SimTrust* and inference algorithms

| Algorithm  | Top 5 | Top 10 | Top 15 | Top 20 | Top 25 | Top 30 |
|------------|-------|--------|--------|--------|--------|--------|
| *DSPG*     | 0.148 | 0.144  | 0.137  | 0.132  | 0.128  | 0.127  |
| *MoleTrust* | 0.110 | 0.103  | 0.097  | 0.093  | 0.089  | 0.087  |
| *TidalTrust* | 0.088 | 0.082  | 0.079  | 0.077  | 0.075  | 0.074  |
| *SimTrust* | 0.065 | 0.056  | 0.052  | 0.048  | 0.043  | 0.039  |

**Table 6.10** Average recall values for *SimTrust* and inference algorithms

| Algorithm | Top 5 | Top 10 | Top 15 | Top 20 | Top 25 | Top 30 |
|-----------|-------|--------|--------|--------|--------|--------|
| *DSPG* | 0.117 | 0.121 | 0.123 | 0.125 | 0.127 | 0.128 |
| *MoleTrust* | 0.077 | 0.083 | 0.086 | 0.090 | 0.093 | 0.099 |
| *TidalTrust* | 0.066 | 0.069 | 0.074 | 0.079 | 0.082 | 0.089 |
| *SimTrust* | 0.031 | 0.047 | 0.061 | 0.072 | 0.078 | 0.084 |

**Table 6.11** Average *F1*-Measure values for *SimTrust* and inference algorithms

| Algorithm | Top 5 | Top 10 | Top 15 | Top 20 | Top 25 | Top 30 |
|-----------|-------|--------|--------|--------|--------|--------|
| *DSPG* | 0.131 | 0.132 | 0.130 | 0.128 | 0.127 | 0.127 |
| *MoleTrust* | 0.091 | 0.092 | 0.091 | 0.091 | 0.091 | 0.093 |
| *TidalTrust* | 0.075 | 0.075 | 0.076 | 0.078 | 0.078 | 0.081 |
| *SimTrust* | 0.042 | 0.051 | 0.056 | 0.058 | 0.055 | 0.053 |

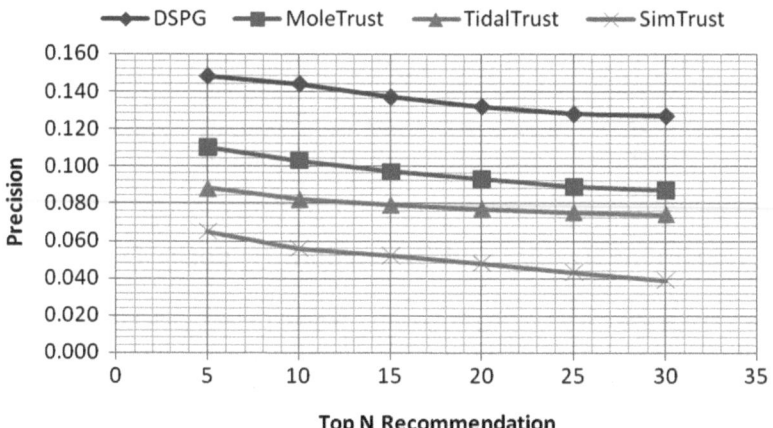

**Fig. 6.9** Evaluation for *Hypothesis 2* and *3* with precision metric

We utilized each of the four techniques to recommend a list of top-*N* items for each of the 3,872 users, and different values for *N* ranging from 5 to 30 were tested. The results of the experiment are shown in Tables 6.9, 6.10 and 6.11 and corresponding graphs in Figure 6.9, 6.10 and 6.11. For this part of the experiment, we used only Approach B for the evaluation as we found the recommendation values are much higher for Approach B than Approach A.

It can be observed from the figures that for all the three evaluation metrics presented in Tables 6.9, 6.10 and 6.11, the proposed *DSPG* inference technique achieved the best result among all the four techniques compared. The two benchmark inference methods *TidalTrust* and *MoleTrust* performed similarly but *DSPG* performed significantly better than both of the methods. All of these three inference methods used the same recommendation techniques but different techniques were used to find neighbors. The results clearly show that when the

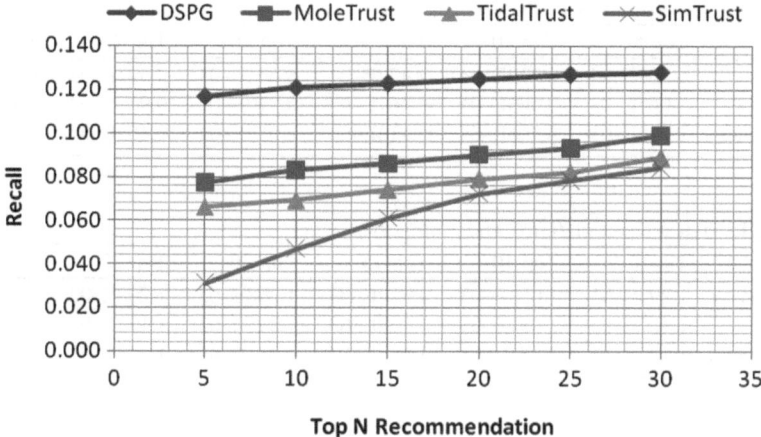

**Fig. 6.10**  Evaluation for *Hypothesis 2* and *3* with recall metric

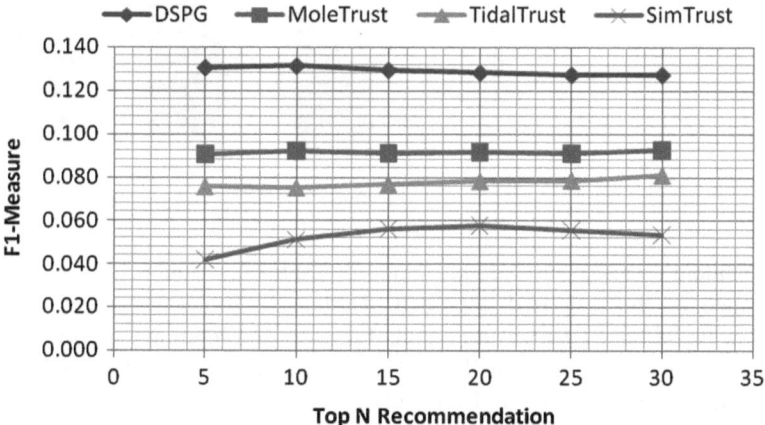

**Fig. 6.11**  Evaluation for Hypothesis 2 and 3 with *F1*-Measure metric

*SimTrust* approach was used directly without the inference, its performance was poor but when the inference technique was used to expand the neighborhoods, the performance improved significantly. It proves our hypothesis 2. Among the three inference techniques compared, our proposed *DSPG* approach outperformed the other two popular inference techniques, which proves that our hypothesis 3 is also correct.

Figure 6.9 shows the precision metric of all four techniques compared. Then, Fig. 6.10 shows the Recall values between the four approaches. And finally in Fig. 6.11, the *F*-Measure based on Precision and Recall of the four approaches is presented.

**Table 6.12** *T*-test result of *DSPG* and *TidalTrust*

|            | Top 5     | Top 10    | Top 15    | Top 20    | Top 25    | Top 30    |
|------------|-----------|-----------|-----------|-----------|-----------|-----------|
| Precision  | 6.809E–04 | 6.220E–04 | 7.477E–04 | 8.662E–04 | 9.597E–04 | 9.597E–04 |
| Recall     | 1.067E–03 | 9.566E–04 | 9.467E–04 | 9.970E–04 | 9.943E–04 | 1.127E–03 |
| *F1*-Measure | 1.020E–03 | 8.746E–04 | 9.113E–04 | 8.952E–04 | 9.467E–04 | 1.061E–03 |

**Table 6.13**  T-test result of *DSPG* and *MoleTrust*

|             | Top 5     | Top 10    | Top 15    | Top 20    | Top 25    | Top 30    |
|-------------|-----------|-----------|-----------|-----------|-----------|-----------|
| Precision   | 1.297E–03 | 1.053E–03 | 1.127E–03 | 1.208E–03 | 1.208E–03 | 1.127E–03 |
| Recall      | 1.127E–03 | 1.051E–03 | 1.127E–03 | 1.045E–03 | 1.127E–03 | 1.127E–03 |
| F1-Measure  | 1.127E–03 | 1.051E–03 | 1.127E–03 | 1.045E–03 | 1.127E–03 | 1.127E–03 |

**Fig. 6.12** Average computation efficiency for different inference techniques

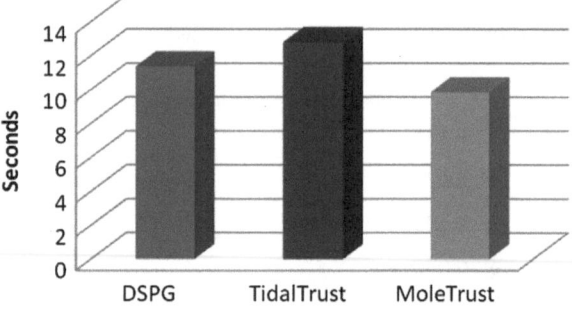

The student's $t$ test results are presented in Tables 6.12 and 6.13 to evaluate the significance of the proposed approach. Table 6.12 shows the $t$ test values of Precision, Recall and *F1*-measure results for proposed inference algorithm *DSPG* and baseline method *TidalTrust*. Table 6.13 shows the similar values of *DSPG* and *MoleTrust*. All the $p$ values of Tables 6.12 and 6.13 are less than 7.000E–02 for Precision, Recall and *F1*-Measure. This means that our proposed approach is significantly better than the baseline models in regards to all of the Precision, Recall and *F1*-Measure values.

The comparison of computation efficiency between the proposed *DSPG* inference algorithm and the two popular *TidalTrust* and *MoleTrust* inference algorithms as a benchmark is shown in Figure 6.12.

**Table 6.14** Processing time comparison for different inference techniques

| Inference algorithm | Processing time |
|---------------------|-----------------|
| *DSPG*              | 12 h 13 m 24 s  |
| *MoleTrust*         | 13 h 46 m 11 s  |
| *TidalTrust*        | 10 h 36 m 52 s  |

The figure shows the average processing time taken for a single user. For the same set of 3,872 test users, the total processing time for *DSPG*, *TidalTrust* and *MoleTrust* is shown in Table 6.14.

It can be seen from the results shown in Fig. 6.12 above that our proposed *DSPG* takes little more time than *MoleTrust* but performs better than *TidalTrust* for processing the experimental dataset. The reason for this is that we have considered all possible paths from the source to the destination to find trusted users, while *TidalTrust* and *MoleTrust* both consider only the shortest and strongest paths. *MoleTrust* controls the path finding via restricting the number of hops by setting a threshold value, which compromises the recommendation quality. As our main focus of this research was to improve the recommendation quality in terms of correct recommendation, we consider the processing time efficiency improvement as a future extension of this work.

## 6.9 Chapter Summary

This chapter has presented the evaluation method and experimental results of evaluating how the tag similarity based trust-network recommendation and the proposed directed serial parallel graph trust transitivity approaches performed. In this chapter, the experimental method with specific hypothesis set for the experiment and details of dataset used are described. It also presented a brief discussion on the parameterization of the variables used for conducting the experiment. Finally, the actual results of the experimental finding are presented with easily understandable graphical forms. The experiment results showed that this tag-based similarity approach performs better while making recommendations than the traditional collaborative filtering based approach. At the end, a comparison of average computation efficiency for different inference techniques has presented.

# Chapter 7
# Conclusions

## 7.1 Contribution

We have presented a new algorithm for generating trust networks based on user tagging information to make recommendations in the online environment. The experiment results showed that this tag-based similarity approach performs better while making recommendations than the traditional collaborative filtering based approach. This proposed technique will be very helpful to deal with data sparsity problems; even when explicit trust rating data is unavailable. The finding will contribute in the area of recommender systems by improving the overall quality of automated recommendation making.

This book contributes to help users solve the information overload issue by proposing a new way of tag-similarity based neighborhood formation for more accurate, effective and efficient recommendation. The contributions of this book can benefit research on recommender systems as well as online trust management. We have showed that commonality of users' tagging behavior and inferred trust relationships offer some real benefits to the user. In order to accomplish this, we provide the following: (1) Use of a trust inference technique to increase the range/boundary of the neighbors and improve recommendation results, (2) Established new knowledge in the area of trust management through a survey on the relationship between trust and interest similarity, (3) Proposed a new trust inference technique which performs better than existing standard trust inference techniques and (4) A new method of trust network generation in the absence of explicit trust rating using web 2.0 information such as "tag" and use it for recommendation making. It makes a number of specific contributions to trust management and recommender systems, which are summarized below:

- Proposed a new trust inference technique which enables superior formation of trust networks for recommendation making (Please refer to Chap. 3). We proposed an algorithm for inferring a trust value from one node to another where no direct connection exists between the nodes in a trust network. Trust Inference can be used to recommend to one node how much to trust another node in the

T. Bhuiyan, *Trust for Intelligent Recommendation,*
SpringerBriefs in Electrical and Computer Engineering,
DOI: 10.1007/978-1-4614-6895-0_7, © The Author(s) 2013

network. In other word, what trust rating one person wants to give another unknown person if there is an indirect link present? Due to the limited number of ratings that exist in recommender systems, underlying networks that are generated based on user rating similarity are very sparse. There are cases in which insufficient or lack of information is detrimental for the recommendation algorithm.

- Established a strong correlation between trust and interest similarity through an online survey (Please refer to Chap. 4). A positive relationship between trust and interest similarity is assumed and adopted by many researchers but no survey work on real life people's opinions to support this hypothesis was found. The survey result supports the assumed hypothesis of a positive relationship between the trust and interest similarity of the users. It is also found that in general, people prefer online recommendation if it comes from a subject expert. But people are uncertain about accepting the people they met online as their friends. We believe that our findings from this survey will have a great impact in the area of recommender system research; especially where discovering user interest similarity plays an important role. The findings of the existence of positive correlation between trust and interest similarity gives us the confidence that the interest similarity value can be used as trust data to build a trust network in the absence of direct explicit trust.

- Proposed a new tag-based algorithm to determine interest similarity to substitute for explicit trust rating information where explicit trust rating information is lacking or unavailable (Please refer to Chap. 5). We proposed an algorithm to generate a trust network in the absence of explicit trust data for generating quality recommendations. The ultimate goal is to create an advanced solution which can improve the automated recommendation making. The existing trust-based recommender works have assumed that the trust network already exists. However, our proposed algorithm performs well even the absence of explicit trust data. To identify the interest similarity, we use user's personalised tagging behaviour information. Our emphasis is on what resources or the contents of those resources the user chooses to tag, rather than the textual name of the tag itself. The commonalities of the resources being tagged by the users' show their common interest and this can be used to form the neighbours used in the automated recommender system.

- Evaluated the impact of inference to increase the boundaries of neighbourhoods for better recommendation using a large real-life dataset (Please refer to Chap. 6). Our experimental results show that our proposed tag-similarity based method outperforms the traditional collaborative filtering approach which usually uses rating data. Thus, we argue that using the proposed approaches, the goal of more accurate, effective and efficient recommender systems can be achieved.

## 7.2 Limitations

The book has some limitations, which are listed below:

- This book did not consider combining tags with other Web 2.0 information such as reviews or opinions, in an attempt to determine whether this would have any positive impact on the recommendation making. Opinion mining could be a useful tool to identify the orientation of the reviews and it could be used to improve the recommendation quality further.
- The experimental data set is taken from an online retail business. It did not verify the effectiveness of the proposed models in a social network environment, which is the largest potential application domain for the findings of this work. A real-life social network dataset could be useful to evaluate the overall performance in a broader domain.
- This book focuses on the improvement of the correctness of recommendations rather than improvements in processing efficiency which is another major concern in recommender systems. Future work should include efforts to improve the algorithms for faster processing especially for a large dataset in a real-life situation.
- The survey conducted for this research to establish the opinion of users on the relationship between trust and interest similarity has used a limited number of respondents. A greater number of participants may increase the acceptance of the findings in the survey.

## 7.3 Future Work

This book can be extended in various directions to overcome the limitations stated in the previous section. Web 2.0 provides additional information such as comments, opinions, reviews etc. in addition to the traditional rating data. Future work could integrate rating data with other available information to improve the recommendation quality. Another approach could be to integrate tag data with other information to find the interest similarity to identify similar users. A larger data set from a real-life social network could be used to evaluate the effectiveness of the proposed method to verify its suitability in the online social network environment. The survey can be extended with more questions and by increasing the number of participants to make it more acceptable. The proposed algorithms can be redesigned to improve processing time to perform more efficiently, thus not only providing correct recommendations but also faster processing.

# Appendix A
# Sample Survey Questions

## About You

Please select the option that best describes your status.

1. Please tell us your age

   - 21–25
   - 26–30
   - 31–40
   - Over 40
   - Don't want to tell

2. Please specify your gender

   - Male
   - Female
   - Don't want to tell

3. Do you have any experience in using online recommender systems?

   - Yes, I've used them many times
   - Yes, very few times
   - No, I've no idea about it

## Your Opinion

Please select the option that best describes your opinion.

4. Do you prefer to have automated recommendation for a product or service?

   - Yes
   - No

T. Bhuiyan, *Trust for Intelligent Recommendation*, 97
SpringerBriefs in Electrical and Computer Engineering,
DOI: 10.1007/978-1-4614-6895-0, © The Author(s) 2013

5. Assume that an unknown automobile expert A and one of your friends B who is not expert about cars is available for recommendation when you are going to buy a car. Which recommendation will you prefer?

   - A
   - B
   - Either one
   - Neither of them

6. Which recommendation will you prefer most?

   - From a friend whom you trust
   - From a person who has similar taste to you

7. Do you consider people you have met online as your friends?

   - Yes, some of them
   - No, it is difficult to trust them

8. Would you like to rate how much you trust your friends?

   - Yes, that would be helpful
   - No, that is not necessary
   - Don't care
   - Undecided

9. If you could rate your online friends, would you be bothered doing so?

   - Yes, I would
   - No, I wouldn't
   - Don't care

10. Which one is more important to you? A recommendation from a person who

   - Has a good reputation
   - Is competent in the area of recommendation
   - Is believed by you
   - deserve confidence

11. In terms of recommendation making, which one best describes your opinion?

   - Automated recommendation generated by an expert system
   - Only from people I know
   - Only from my family and friends

12. One of your friends X has similar taste to you when selecting movies and another friend Y is interested in different types of movies.

   - I will trust X more than Y to make a movie recommendation for me
   - Either one is equal to me as long as I know them
   - Neither of them

13. Do you think there is any relationship between "Trust" and "Interest Similarity"?

- Yes
- No
- Maybe

# Appendix B
# Glossary of Terms and Abbreviations

| | |
|---|---|
| **B2B** | Business to Business |
| **BFS** | Breadth First Search |
| **CF** | Collaborative Filtering |
| **CPU** | Central Processing Unit |
| **DBMS** | Data Base Management System |
| **DFS** | Depth First Search |
| **DSPG** | Directed Series Parallel Graph |
| **ERA** | Excellence in Research for Australia |
| **GB** | Giga Byte |
| **GH** | Giga Hertz |
| **HDR** | Higher Degree by Research |
| **HPC** | High Performance Computer |
| **HREC** | Human Research Ethics Committee |
| **IDB** | Information Database |
| **IMDb** | Internet Movie Database |
| **ISBN** | International Standard Book Number |
| **LDA** | Latent Dirichlet Allocation |
| **MAU** | Multi-Attribute Utility |
| **OBD** | Outcomes Database |
| **PLSI** | Probabilistic Latent Semantic Index |
| **QoS** | Quality of Service |
| **QUT** | Queensland University of Technology |
| **SDB** | Sociograms Database |
| **SQL** | Structured Query Language |
| **TF-IDF** | Term Frequency/Inverse Document Frequency |
| **TF-IUF** | Term Frequency/Inverse User Frequency |
| **TNA-SL** | Trust Network Analysis with Subjective Logic |
| **UGC** | User Generated Content |

T. Bhuiyan, *Trust for Intelligent Recommendation*,
SpringerBriefs in Electrical and Computer Engineering,
DOI: 10.1007/978-1-4614-6895-0, © The Author(s) 2013

Appendix B
Glossary of Terms and Abbreviations

# Appendix C
# Combining Trust and Reputation Management

Services offered and provided through the Web have varying quality, and it is often difficult to assess the quality of a service before accessing and using it. Trust and reputation systems can be used in order to assist users in predicting and selecting the best quality services. This section describes how Bayesian reputation systems can be combined with trust modeling based on subjective logic to provide an integrated method for assessing the quality of online services. This will not only assist the user's decision making, but will also provide an incentive for service providers to maintain high quality, and can be used as a sanctioning mechanism to discourage deceptive and low quality services.

Online trust and reputation systems are emerging as important decision support tools for selecting online services and for assessing the risk of accessing them. Binomial Bayesian reputation systems normally take ratings expressed in a discrete binary form as either positive (e.g. *good*) or negative (e.g. *bad*). Multinomial Bayesian reputation systems allow the possibility of providing ratings with discrete graded levels such as for example, *mediocre–bad–average–good–excellent*. It is also possible to use continuous ratings in both binomial and multinomial reputation systems. Multinomial models have the advantage that scores can distinguish between the case of polarized ratings (e.g. a combination of strictly good and bad ratings) and the case of only average ratings. Trust models based on subjective logic are directly compatible with Bayesian reputation systems because a bijective mapping exists between their respective trust and reputation representations. This provides a powerful basis for combining trust and reputation systems for assessing the quality of online services.

A general characteristic of reputation systems is that they provide global reputation scores, which means that all the members in a community will see the same reputation score for a particular agent. On the other hand, trust systems can in general be used to derive local and subjective measures of trust. Different agents can derive different trust values for the same entity. Another characteristic of trust systems is that they can analyze multiple hops of trust transitivity. Reputation systems on the other hand normally compute scores based on direct input from members in the community and not based on transitivity. Still there are systems

T. Bhuiyan, *Trust for Intelligent Recommendation*,
SpringerBriefs in Electrical and Computer Engineering,
DOI: 10.1007/978-1-4614-6895-0, © The Author(s) 2013

**Table C.1** Possible combinations of local/global scores and transitivity/no transitivity

|                 | Private scores                                      | Public scores                            |
| --------------- | --------------------------------------------------- | ---------------------------------------- |
| Transitivity    | Trust systems<br>e.g. Rummble.com                   | Public trust systems<br>e.g. *Page Rank* |
| No Transitivity | Private reputation systems<br>e.g. customer feedback analysis | Reputation systems<br>e.g. eBay.com |

that have characteristics of being both a reputation system and a trust system. The matrix below shows examples of the possible combinations of local and global scores, and trust transitivity or not.

In this section a framework for combining these forms of trust and reputation systems has been described. Because Bayesian reputation systems are directly compatible with trust systems based on subjective logic, they can be seamlessly integrated. This provides a powerful and flexible basis for online trust and reputation management.

## The Dirichlet Reputation System

Reputation systems collect ratings about users or service providers from members in a community. The reputation centre is then able to compute and publish reputation scores about those users and services. Figure C.1 illustrates a reputation centre where the dotted arrow indicates ratings and the solid arrows indicate reputation scores about the users.

Multinomial Bayesian systems are based on computing reputation scores by statistical updating of Dirichlet Probability Density Functions (PDF), which therefore are called Dirichlet reputation systems. A *posteriori* (i.e. the updated) reputation scores are computed by combining a priori (i.e. previous) reputation scores with new ratings.

In Dirichlet reputation systems, agents are allowed to rate others agents or services with any level from a set of predefined rating levels, and the reputation scores are not static but will gradually change with time as a function of the received ratings. Initially, each agent's reputation is defined by the base rate reputation, which is the same for all agents. After ratings about a particular agent are received, that agent's reputation will change accordingly.

Let there be $k\psi$ different discrete rating levels. This translates into having a state space of cardinality $k\psi$ for the Dirichlet distribution. Let the rating level be indexed by $i$. The aggregate ratings for a particular agent are stored as a cumulative vector, expressed as:

**Fig. C.1** Simple reputation system

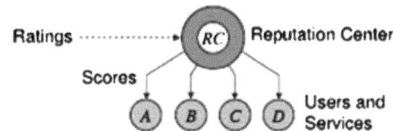

$$\overrightarrow{R} = (\overrightarrow{R}(L_i)|i = 1\ldots k). \tag{C.1}$$

This vector can be computed recursively and can take factors such as longevity and community base rate into account. The most direct form of representing a reputation score is simply the aggregate rating vector $\overrightarrow{R}_y$ which represents all relevant previous ratings. The aggregate rating of a particular level $i$ for agent $y$ is denoted by $\overrightarrow{R}_y(L_i)$.

For visualization of reputation scores, the most natural is to define the reputation score as a function of the probability expectation values of each rating level. Before any ratings about a particular agent y have been received, its reputation is defined by the common base rate $\overrightarrow{a}$. As ratings about a particular agent are collected, the aggregate ratings can be computed recursively and the derived reputation scores will change accordingly. Let $\overrightarrow{R}$ represent a target agent's aggregate ratings. Then the vector $\overrightarrow{S}$ defined by:

$$\overrightarrow{S}_y : \left( \overrightarrow{S}_y(L_i) = \frac{\overrightarrow{R}_y(L_i) + C\overrightarrow{a}(L_i)}{C + \sum_{j=1}^{k} \overrightarrow{R}_y(L_j)} ; |i = 1\ldots k \right). \tag{C.2}$$

is the corresponding multinomial probability reputation score. The parameter $C$ represents the non-informative prior weight where $C = 2$ is the value of choice, but a larger value for the constant $C$ can be chosen if a reduced influence of new evidence over the base rate is required.

The reputation score $\overrightarrow{S}$ can be interpreted like a multinomial probability measure as an indication of how a particular agent is expected to behave in future transactions. It can easily be verified that:

$$\sum_{i=1}^{k} \overrightarrow{S}(L_i) = 1. \tag{C.3}$$

While informative, the multinomial probability representation can require considerable space on a computer screen because multiple values must be visualized. A more compact form can be to express the reputation score as a single value in some predefined interval. This can be done by assigning a point value $v$ to each rating level $L_i$, and computing the normalized weighted point estimate score $\sigma$.

Assume, for example, $k$ different rating levels with point values $v(L_i)$ evenly distributed in the range $[0, 1]$ according to $v(L_i) = \frac{i-1}{k-1}$. The point estimate reputation score of a reputation $\overrightarrow{R}$ is then:

$$\sigma = \sum_{i=1}^{k} v(L_i) \overrightarrow{S}(L_i). \tag{C.4}$$

A point estimate in the range $[0, 1]$ can be mapped to any range, such as 1–5 stars, a percentage or a probability.

**Fig. C.2** Scores and point
estimate during a sequence of
varying ratings

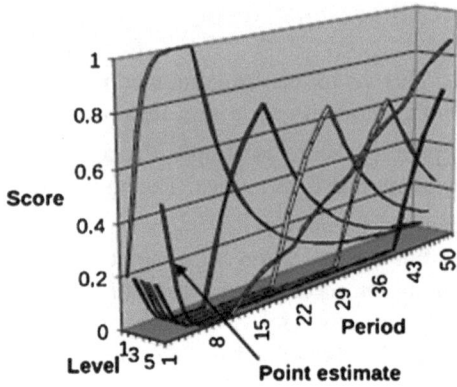

Bootstrapping a reputation system to a stable and conservative state is important. In the framework described above, the base rate distribution $\overrightarrow{a}$ will define an initial default reputation for all agents. The base rate can, for example, be evenly distributed over all rating levels, or biased towards either negative or positive rating levels. This must be defined when setting up the reputation system in a specific market or community.

As an example we consider five discrete rating levels, and the following sequence of ratings:

- Periods 1–10: L1 Mediocre
- Periods 11–20: L2 Bad
- Periods 21–30: L3 Average
- Periods 31–40: L4 Good
- Periods 41–50: L5 Excellent

The longevity factor is $\lambda = 0.9$, and the base rate is dynamic. The evolution of the scores of each level as well as the point estimate is illustrated in Fig. C.2.

In Fig. C.2 the multinomial reputation scores change abruptly between each sequence of 10 periods. The point estimate first drops as the score for L1 increases during the first 10 periods. After that the point estimate increases relatively smoothly during the subsequent 40 periods. Assuming a dynamic base rate and an indefinite series of L5 (Excellent) ratings, the point estimate will eventually converge to 1.

## Proposed Method for Combining

A bijective mapping can be defined between multinomial reputation scores and opinions, which makes it possible to interpret these two mathematical representations as equivalent. The mapping can symbolically be expressed as:

$$\omega \leftrightarrow \overrightarrow{R}. \tag{C.5}$$

This equivalence can be expressed as (Jøsang 2007):

**Theorem 1** (Equivalence between opinions and reputations)
*Let $\omega = (\overrightarrow{b}, u, \overrightarrow{a})$ be an opinion, and $\overrightarrow{R}$ be a reputation, both over the same state space X so that the base rate $\overrightarrow{a}$ also applies to the reputation. Then the following equivalence holds (Jøsang 2007):*
*For $u \neq 0$ :*

$$
\begin{cases}
\overrightarrow{b}(x_i) = \dfrac{\overrightarrow{R}(x_i)}{C + \sum_{i=1}^{k} \overrightarrow{R}(x_i)} \\[3mm]
u = \dfrac{C}{C + \sum_{i=1}^{k} \overrightarrow{R}(x_i)}
\end{cases}
\Leftrightarrow
\begin{cases}
\overrightarrow{R}(x_i) = \dfrac{c\,\overrightarrow{b}(x_i)}{u} \\[3mm]
u + \sum_{i=1}^{k} \overrightarrow{b}(x_i) = 1
\end{cases}
\tag{C.6}
$$

*For $u = 0$ :*

$$
\begin{cases}
\overrightarrow{b}(x_i) = \eta(x_i) \\
u = 0
\end{cases}
\Leftrightarrow
\begin{cases}
\overrightarrow{R}(x_i) = \eta(x_i) \sum_{i=1}^{k} \overrightarrow{R}(x_i) = \eta(x_i)\infty \\[3mm]
\sum_{i=1}^{k} m(x_i) = 1
\end{cases}
\tag{C.7}
$$

The case of $u = 0$ reflects an infinite amount of aggregate ratings, in which case the parameter $\eta$ determines the relative proportion of infinite ratings among the rating levels. In case $u = 0$ and $\eta(x_i) = 1$ for a particular rating level $x_i$, then $\overrightarrow{R}(x_i) = \infty$ and all the other rating parameters are finite. In the case of $\eta(x_i) - 1/k$ for all $i = 1...k$, then all the rating parameters are equally infinite. As already indicated, the non-informative prior weight is normally set to $C = 2$.

Multinomial aggregate ratings can be used to derive binomial trust in the form of an opinion. This is done by first converting the multinomial ratings to binomial ratings according to Eq. (C.8) below, and then to apply Theorem 1.

Let the multinomial reputation model have k rating levels $x_i$; $i = 1,...k$, where $\overrightarrow{R}(x_i)$ represents the ratings on each level $x_i$, and let $\sigma$ represent the point estimate reputation score from Eq. (C.4). Let the binomial reputation model have positive and negative ratings $r$ and $s$ respectively. The derived converted binomial rating parameters $(r, s)$ are given by:

$$
\begin{cases}
r = \sigma \sum_{i=1}^{k} \overrightarrow{R}_y(x_i) \\
s = \sum_{i=1}^{k} \overrightarrow{R}_y(x_i) - r
\end{cases}
\tag{C.8}
$$

With the equivalence of Theorem 1 it is possible to analyze trust networks based on both trust relationships and reputation scores. Figure C.3 illustrates a scenario where agent $A$ needs to derive a measure of trust in agent $F$.

**Fig. C.3** Combining trust
and reputation

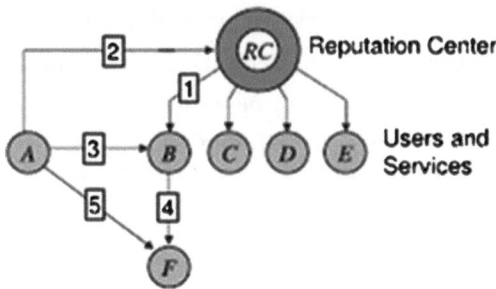

Agent $B$ has reputation score $\overrightarrow{R}_B^{RC}$ (arrow 1), and agent $A$ has trust $\omega_{RC}^A$ in the Reputation Centre (arrow 2), so that $A$ can derive a measure of trust in $B$ (arrow 3). Agent $B$'s trust in $F$ (arrow 4) can be recommended to $A$ so that $A$ can derive a measure of trust in $F$ (arrow 5). Mathematically this can be expressed as:

$$\omega_F^A = \omega_{RC}^A \otimes \overrightarrow{R}_B^{RC} \otimes \omega_F^B \tag{C.9}$$

The compatibility between Bayesian reputation systems and subjective logic makes this a very flexible framework for analyzing trust in a network consisting of both reputation scores and private trust values.

## Trust Derivation Based on Trust Comparisons

It is possible that different agents have different trust in the same entity, which intuitively could affect the mutual trust between the two agents. Figure C.4 illustrates a scenario where $A$'s trust $\omega_B^A$ (arrow 1) conflicts with $B$'s reputation score $\vec{R}_B^{RC}$ (arrow 2).

As a result, $A$ will derive a reduced trust value in the Reputation Centre (arrow 3). Assume that $A$ needs to derive a trust value in $E$, then the reduced trust value must be taken into account when using $RC$'s reputation score for computing trust in $E$. The operator for deriving trust based on trust conflict produces a binomial opinion over the binary state space $\{x, \bar{x}\}$, where $x$ is a proposition that can be interpreted as $x$: $RC$ *provides reliable reputation scores*, and $\bar{x}$ is its complement. Binomial opinions

**Fig. C.4** Deriving trust from
conflicting trust

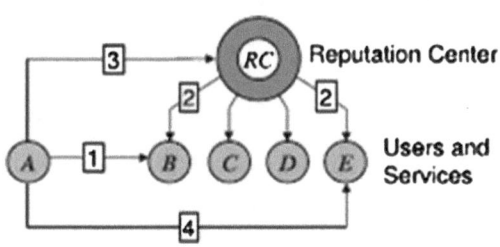

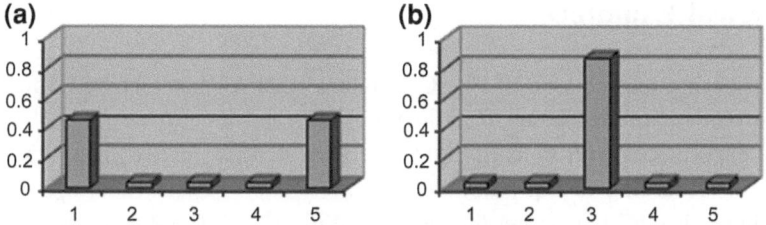

**Fig. C.5** Comparison of polarized and average reputation scores

have the special notation $\omega_x = (b, d, u, a)$ where $d$ represents disbelief in proposition $x$.

What represents difference in trust values depends on the semantics of the state space. Assume that the state space consists of five rating levels, then Fig. C.5a represents a case of polarized ratings, whereas Fig. C.5b represents a case of average ratings. Interestingly they have the same point estimate of 0.5 when computed with Eq. (C.4).

We will define an operator which derives trust based on point estimates as defined by Eq. (C.3). Two agents having similar point estimates about the same agent or proposition should induce mutual trust, and dissimilar point estimates should induce mutual distrust.

**Definition C.1** (*Trust Derivation Based on Trust Comparison*)
Let $\omega_B^A$ and $\omega_B^{RC}$ be two opinions on the same state space B with a set rating level. A's trust in RC based on the similarity between their opinions is defined as:

$$\omega_{RC}^A = \omega_B^A \downarrow \omega_B^{RC} \ where \begin{cases} d_{RC}^A = |\sigma(\vec{R}_B^A) - \sigma(\vec{R}_B^{RC})| \\ u_{RC}^A = Max[u_B^A, u_B^{RC}] \\ b_{RC}^A = 1 - b_{RC}^A - u_{RC}^A \end{cases} \qquad (C.10)$$

The interpretation of this operator is that disbelief in $RC$ is proportional to the greatest difference in point estimates between the two opinions. Also, the uncertainty is equal to the greatest uncertainty of the two opinions.

With the trust comparison trust derivation operator, A is able to derive trust in $RC$ (arrow 3). With the above described trust and reputation measures, A is able to derive trust in $E$ expressed as:

$$\omega_E^A = \omega_{RC}^A \otimes \omega_E^{RC} \qquad (C.11)$$

This provides a method for making trust derivation more robust against unreliable or deceptive reputation scores and trust recommendations.

## Numerical Examples

By considering the scenario of Fig. C.4, assume that $RC$ has received 5 mediocre and 5 excellent ratings about $B$ as in Fig. C.5a, and that $A$ has had 10 average private experiences with $B$, as in Fig. C.5b. Then $\sigma(\vec{R}_B^{RC}) = \sigma(\vec{R}_B^A) = 0.5$, so that $d_{RC}^A = 0$. According to Eq. (C.6) we get $u_B^{RC} = u_B^A = 1/6$, so that $u_{RC}^A = \frac{1}{6}$ and according to Eq. (C.10) we get $b_{RC}^A = 5/6$. With $a_{RC}^A = 0.9$ the derived binomial opinion is $\omega_{RC}^A = (5/6, 0, 1/6, 0.9)$, which indicates a relatively strong, but somewhat uncertain trust.

Assume further the aggregate ratings $\vec{R}_E^{RC} = (0, 4, 2, 2, 0)$, i.e. reflecting 0 mediocre, 4 bad, 2 average, 2 good and 0 excellent ratings about $E$. The base rate vector is set to $\vec{a} = (0.1, 0.2, 0, 2, 0.4, 0.1)$ and the non-informative prior weight $C = 2$. Using Eq. (C.2), the multinomial reputation score is $\vec{S}_E = (0.02, 0.44, 0.24, 0.28, 0.02)$. The point values for each level from mediocre to excellent are: 0.00, 0.25, 0.50, 0.75 and 1.00. Using Eq. (C.4) the point estimate reputation is $\sigma = 0.46$.

Using Eq. (C.8) and the fact that $\sum_{i=1}^k \vec{R}_E^{RC}(x_i) = 8$, the reputation parameters can be converted to the binomial $(r, s) = (3.68, 4.32)$. Using Eq. (C.6), $RC$'s trust in $E$ in the form of a binomial opinion can be computed as $\omega_E^{RC} = (0.368, 0.432, 0.200, 0.500)$ where the base rate trust has been set to $a_E^{RC} = 0.5$.

The transitivity operator can now be used to derive $A$'s trust in $E$. The base rate sensitive operator can be used, which for this example is expressed as:

$$\begin{cases} b_E^{A:RC} = (b_{RC}^A + a_{RC}^A u_{RC}^A) b_E^{RC} \\ d_E^{A:RC} = (b_{RC}^A + a_{RC}^A u_E^A) d_E^{RC} \\ u_E^{A:RC} = 1 - b_E^{A:RC} - d_E^{A:RC} \\ a_E^{A:RC} = a_E^{RC} \end{cases} \quad (C.12)$$

$A$'s trust in $E$ can then be computed as the opinion $\omega_E^A = (0.362, 0.425, 0.213, 0.500)$, which in terms of probability expectation value is $E(\omega_E^A) = 0.4686$. This rather weak trust was to be expected from the relatively negative ratings about $E$.

## Conclusions

Trust and reputation management represents an important approach for stabilizing and moderating online markets and communities. Integration of different systems would be problematic with incompatible trust and reputation systems. We have described how it is possible to elegantly integrate Bayesian reputation systems and trust analysis based on subjective logic. This provides a flexible and powerful framework for online trust and reputation management.

# References

Abdul-Rahman, A., & Hailes, S. (2000). Supporting trust in virtual communities. *Proceedings of the 33rd Hawaii International Conference on System Sciences,* Hawaii, HI.

Adomavicius, G., & Tuzhilin, A. (2005). Toward the next generation of recommender systems: a survey of the state-of-the-art and possible extensions. *IEEE Transactions on Knowledge and Data Engineering, 17*(6), 734–749.

Arapakis, I., Moshfeghi, Y., Joho, H., Ren, R., Hannah, D. & Jose, J. M. (2009). Enriching user profiling with affective features for the improvement of a multimodal recommender system. *Proceedings of the ACM International Conference on Image and Video Retrieval, Island of Santorini, Greece.*

Avesani, P., Massa, P. & Tiella, R. (2004). Mole skiing: trust-aware decentralized recommender system. *1st Workshop on Friend of a Friend, Social Networking and the Semantic Web.* Galway, Ireland.

Avesani, P., Massa, P., & Tiella, R. (2005). A trust-enhanced recommender system application: Mole skiing. *Proceedings of the 2005 ACM Symposium on Applied Computing* (pp. 1589–1593). New York

Badrul, S., George, K. Joseph, K., & John, R. (2001). Item-based collaborative filtering recommendation algorithms. *Proceedings of the 10th International Conference on World Wide Web,* Hong Kong, China.

Bao, S., Wu, X., Fei, B., Xue, G., Su, Z., & Yu, Y. (2007). Optimizing Web Search Using Social Annotations. *Proceedings of the 16th International Conference on World Wide Web* (pp. 501–510).

Basu, C., Hirsh, H., & Cohen, W. W. (1998). Recommendation as classification: Using social and content-based information in recommendation. *Proceedings of the 15th National Conference on Artificial Intelligence* (pp. 714–720).

Balabanovi, M., & Shoham, Y. (1997). *Fab*: Content-based, collaborative recommendation. *ACM Communications, 40*(3), 66–72.

Bedi, P., & Kaur, H. (2006). Trust based personalized recommender system. *INFOCOM Journal of Computer Science, 5*(1), 19–26.

Bedi, P., Kaur, H., & Marwaha, S. (2007). Trust based recommender system for the semantic web. *Proceedings of the IJCAI07, Hyderabad, India* (pp. 2677–2682).

Begelman, G., Keller, P., & Smadja, F. (2006). Automated tag clustering: Improving search and exploration in the tag space. *Proceedings of the Collaborative Web Tagging Workshop of WWW 2006.*

Berscheid, E. (1998). Interpersonal attraction. In D. Gilbert, S. Fiske & G. Lindzey (Eds.), *The handbook of social psychology* (4th ed., Vol. 2). New York: McGraw-Hill

Bhargava, H. K., Sridhar, S., & Herrick, C. (1999). Beyond spreadsheets: Tools for building decision support systems. *IEEE Computer, 32*(3), 31–39.

Billsus, D., & Pazzani, M. (1999). A hybrid user model for news classification. *Proceedings of the 7th International Conference on User Modelling* (pp. 99–108).

Bindelli, S., Criscione, C., Curino, C., Drago, M.L., Eynard, D., & Orsi, G. (2008). Improving search and navigation by combining ontologies and social tags. *Proceedings of the 2008 OTM Confederated International Workshops* (pp. 76–85).

Bogers, T., & Bosch, A. (2008). Recommending scientific articles using CiteULike. *Proceedings of the ACM Conference on Recommender Systems* (pp. 287–290).

Breese, J. S., Heckerman, D., & Kadie, C. (1998). Empirical analysis of predictive algorithms for collaborative filtering. *Proceedings of the 14th Conference on Uncertainty in Artificial Intelligence, Madison, WI.*

Brin, S., & Page, L. (1998). The anatomy of a large-scale hyper textual Web search engine. *Computer Networks and ISDN Systems, 30*(1–7), 107–117.

Burgess, E., & Wallin, P. (1943). Homogamy in Social Characteristics. *American Journal of Sociology, 49*(2), 117–124.

Burke, R. (2002). Hybrid recommender systems: Survey and experiments. *User Modeling and User-Adapted Interaction, 12*(4), 331–370.

Burke, R., Hammond, K., & Young, B. (1997). The find me approach to assisted browsing. *IEEE Expert, 12*(4), 32–40.

Byrne, D. (1961). Interpersonal attraction and attitude similarity. *Journal of Abnormal and Social Psychology, 62*(3), 713–715.

Byrne, D. (1971). *The attraction paradigm.* USA: Academic Press.

Chen, Z., Cao, J., Song, Y., Guo, J., Zhang, Y, & Li, J. (2010). Context-oriented web video tag recommendation. *Proceedings of the 19th International Conference on World Wide Web, Raleigh, NC* (pp. 1079–1080).

Choi, S. H., & Cho, Y. H. (2004). An utility range-based similar product recommendation algorithm for collaborative companies. *Expert Systems with Applications, 27*(4), 549–557.

Coster, R., Gustavsson, A., Olsson, T., & Rudstrom, A. (2002). Enhancing web-based configuration with recommendations and cluster-based help. *Proceedings of the workshop on Recommendation and Personalization in e-Commerce, Malaga, Spain*

Coughlan, M., Cronin, P., & Ryan, F. (2009). Survey research: Pprocess and limitations. *International Journal of Therapy and Rehabilitation, 16*(1), 9–15.

Cleverdon, C. W., Mills, J., & Keen, M. (1966). Factors determining the performance of indexing systems. ASLIB Cranfield Project, Cranfield

Czaja, R., & Blair, J. (2005). *Designing surveys: A guide to decisions and procedures.* Thousand Oaks, CA: Pine Forge Press.

Dasgupta, P. (1990). Trust as a commodity. In D. Gambetta (Ed.), *Trust: Making and breaking cooperative relations.* Oxford: Basil Blackwell.

Deshpande, M., & Karypis, G. (2004). Item-based top-N recommendation algorithms. *ACM Transactions on Information Systems, 22,* 143–177.

Deutsch, M. (2004). *Distributive justice: A social psychological perspective.* USA: Yale University Press.

Dimitrakos, T. (2003). A Service-oriented trust management framework. *International Workshop on Deception, Fraud and Trust in Agent Societies, Bologna, Italy* (pp. 53–72).

DiNucci, D. (1999). Fragmented Future. *Print, 53*(4), 20–22.

Efron, M. (2010) Hashtag retrieval in a micro blogging environment. *Proceedings of the 33rd International ACM SIGIR Conference on Research and Development in Information Retrieval, Geneva, Switzerland* (pp. 787–788).

Esfandiari, B., & Chandrasekharan, S. (2001). On how agents make friends: mechanisms for trust acquisition. *Proceedings of the Fifth International Conference on Autonomous Agents Workshop on Deception, Fraud and Trust in Agent Societies.*

Etzioni, A. (1996). A moderate communitarian proposal. *Political Theory, 24*(2), 155–171.

Ferman, A. M., Errico, J. H., Beek, P. V., & Sezan, M. I. (2002). Content-based filtering and personalization using structured metadata. *Proceedings of the 2nd ACM/IEEE-CS joint Conference on Digital Libraries, Portland, Oregon, USA.*

Fitting, M. (1994). Kleene's three-valued logics and their children. *Fundamenta Informaticae, 20,* 113–131.

Folkerts, J. B. (2005). A comparison of reputation-based trust systems. Dissertation M.Sc., Rochester Institute of Technology, USA.

Fu, W.-T., Kannampallil, T. G., & Kang, R. (2009). A semantic imitation model of social tag choices. *Proceedings of the IEEE Conference on Social Computing, Vancouver, BC* (pp. 66–72).

Gal-Oz, N., Gudes, E., & Hendler, D. (2008). A robust and knot-aware trust-based reputation model. *Proceedings of the joint iTrust and PST Conference on Privacy, Trust Management and Security* (pp. 167–182). Berlin: Springer.

Gambetta, D. (Ed.). (2000). *Can we trust trust?* (Vol. 13). Oxford: University of Oxford

Gemmis, M. D., Lops, P., Semeraro, G., & Basile, P. (2008). Integrating tags in a semantic content-based recommender. *Proceedings of ACM Conference on Recommender Systems* (163–170).

Golbeck, J. (2005). Computing and applying trust in web-based social networks. Doctoral Dissertation, University of Maryland College Park, USA.

Golbeck, J. (2006). Combining provenance with trust in social networks for semantic web content filtering. *Proceedings of the International Provenance and Annotation Workshop, Chicago, Illinois, USA* (Vol. 6, pp. 101–108).

Golbeck, J. (2007). The dynamics of web-based social networks: Membership, relationships and change. *First Monday, 12*(11), .

Golbeck, J. (2009). Trust and nuance profile similarity in online social network. *ACM Transactions on the Web, 3*(4), 12.1–33.

Golbeck, J., & Hendler, J. (2006). Inferring binary trust relationships in Web-based social networks. *ACM Transactions on Internet Technology, 6*(4), 497–529.

Goldberg, D., Nichols, D., Oki, B. M., & Terry, D. (1992). Using collaborative filtering to weave an information tapestry. *Communications of the ACM, 35,* 61–70.

Golder, S., & Huberman, B. A. (2005). The structure of collaborative tagging systems. Technical Report, HP Labs

Guan, S., Ngoo, C. S., & Zhu, F. (2002). Handy broker: An intelligent product-brokering agent for m-commerce applications with user preference tracking. *Electronic Commerce Research and Applications, 1*(3–4), 314–330.

Guha, R. V., Kumar, R., Raghavan, P., & Tomkins, A. (2004). Propagation of trust and distrust. *The Proceedings of the 13th International World Wide Web Conference* (pp. 403–412).

Gui-Rong, X., Chenxi, L., Qiang, Y., Wensi, X., Hua-Jun, Z., Yong, Y., & Zheng, C. (2005). Scalable collaborative filtering using cluster-based smoothing. *Proceedings of the 28th Annual International ACM SIGIR Conference on Research and Development in Information Retrieval, Salvador, Brazil* (pp. 114–121).

Guttman, R. H., Moukas, A. G., & Maes, P. (1998). Agent-mediated electronic commerce: A survey. *Knowledge Engineering Review, 13*(2), 147–159.

Hagen, P., Manning, H., & Souza, R. (1999). *Smart Personalization.* Cambridge, MA: Forrester Research.

Halpin, H., Robu, V., & Shepherd, H. (2007). The complex dynamics of collaborative tagging. *Proceedings of the 16th International Conference on World Wide Web* (pp. 211–220).

Harding, R., & Peel, E. (2007). Surveying sexualities: Internet research with non-heterosexuals. *Fem Phychol, 17,* 277–285.

Hayes, C., Avesani, P., & Veeramachanei, S. (2007). An analysis of the use of tags in a blog recommender system. *Proceedings of the 20th International Joint Conference on Artificial Intelligence, Hyderabad, India* (pp. 2772–2777).

Herlocker, J. L., Konstan, J. A., Terveen, L. G., & Riedl, J. T. (2004). Evaluating collaborative filtering recommender systems. *ACM Transactions on Information Systems, 22,* 5–53.

Heymann, P., Ramage, D., & Garcia-Molina, H. (2008). Social tag prediction. *Proceedings of the 31st Annual International ACM SIGIR Conference on Research and Development in Information Retrieval, New York* (pp. 531–538).

Hill, W., Stead, L., Rosenstein, M., & Furnas, G. (1995). Recommending and evaluating choices in a virtual community of use. *Proceedings of the SIGCHI Conference on Human Factors in Computing Systems, New York* (pp. 194–201).

Huang, J., Thornton, K. M., & Efthimiadis, E. N. (2010). Conversational tagging in twitter. *Proceedings of the 21st ACM Conference on Hypertext and Hypermedia, Toronto, ON* (pp. 173–178).

Hussain, F. K., & Chang, E. (2007). An overview of the interpretations of trust and reputation. *The Third Advanced International Conference on Telecommunications*

Hwang, C., & Chen, Y. (2007). Using trust in collaborative filtering recommendation. *Lecture Notes in Computer Science, 4570*, 1052–1060.

Innovation Network (2009). *Data collection tips: Developing a survey.* Retrieved November 1, 2009, from www.innonet.org/client_docs/File/Survey_Dev_Tips.pdf.

Jamali, M., & Ester M. (2009). Trust walker: A random walk model for combining trust-based and item-based recommendation. *Proceedings of the 15th ACM Conference on Knowledge Discovery and Data mining.*

Jaschke, R., Marinho, L., Hotho, A., Schmidt-Thieme, L., & Stumme, G. (2007). Tag recommendations in folksonomies. *Proceedings of the 11th European Conference on Principles and Practice of Knowledge Discovery in Databases* (pp. 506–514).

Jensen, C., Davis, J., & Farnham, S. (2002). Finding others online: Reputation systems for social online spaces. *Proceedings of the Conference on Human Factors in Computing System* (447–454).

Jian, C., Jian, Y., & Jin, H. (2005). Automatic content-based recommendation in e-commence. *Proceedings of the IEEE International Conference on e-Technology, e-Commerce and e-Service* (pp. 748–753). Washington, USA.

Jøsang, A. (2001). A logic for uncertain probabilities. *International Journal of Uncertain, Fuzziness and Knowledge-based Systems, 9*(3), 279–311.

Jøsang, A. (2007). Probabilistic logic under uncertainty. *Proceedings of Computing: The Australian Theory Symposium, 65,* Ballarat, Australia.

Jøsang, A. (2009). Fission of opinions in subjective logic. *Proceedings of the 12th International Conference on Information Fusion*, Seattle, WA, USA.

Jøsang, A. (2010). *Subjective Logic.* Retrieved November 1, 2010, from http://persons.unik.no/josang/papers/ subjective_logic.pdf.

Jøsang, A., & McAnally, S. (2004). Multiplication and comultiplication of belief. *International Journal of Approximate Reasoning, 38*(1), 19–51.

Jøsang, A., & Pope, S. (2005). Semantic constraints for trust transitivity. *Proceedings of the Asia-Pacific Conference of Conceptual Modelling, 43.*

Jøsang, A. Hayward, R. & Pope, S. (2006). Trust Network Analysis with Subjective Logic. *Proceedings of the 29th Australasian Computer Science Conference*, (CRPIT Vol. 48), Hobart, Australia.

Jøsang, A., Ismail, R., & Boyd, C. (2007). A survey of trust and reputation systems for online service provision. *Decision Support Systems, 43*(2), 618–644.

Jun, W., Arjen, P. D. V., & Marcel, J. T. R. (2006). Unifying user-based and item-based collaborative filtering approaches by similarity fusion. *Proceedings of the 29th Annual International ACM SIGIR Conference on Research and Development in Information Retrieval* (pp. 501–508). Seattle, WA.

Karypis, G. (2001). Evaluation of item-based top-N recommendation algorithms. *Proceedings of the 10th Conference of Information and Knowledge Management* (pp. 247–254).

Kazeniac, A. (2009). *Social networks: Face book takes over top spot, twitter climbs.* Retrieved February 10, 2010, from http://blog.compete.com/2009/02/09 /facebook-myspace-twitter-social-network/.

Keser, C. (2003). Experimental games for the design of reputation management systems. *IBM Systems Journal, 42*(3), 498–506.

Konstan, J. A., Miller, B. N., Maltz, D., Herlocker, J. L., Gordon, L. R., & Riedl, J. (1997). Group lens: Applying collaborative filtering to Usenet news. *Communications of the ACM, 40*(3), 77–87.

Krulwich, B. (1997). Life style finder: Intelligent user profiling using large-scale demographic data. *AI Magazine, 18*(2), 37–45.

Lathia, N., Hailes, S & Capra, L. (2008). Trust-based collaborative filtering. *Proceedings of the joint iTrust and PST Conference on Privacy, Trust Management and Security* (pp. 119–134). Berlin: Springer.

Lee, W.-P. (2004). Towards agent-based decision making in the electronic marketplace: Interactive recommendation and automated negotiation. *Expert Systems with Applications, 27*(4), 665–679.

Lesani, M., & Montazeri, N. (2009). Fuzzy trust aggregation and personalized trust inference in virtual social networks. *Computational Intelligence, 25*, 51–83.

Levien, R. (2004). *Attack-resistant trust metrics*. Doctoral Dissertation, University of California, Berkeley, CA, USA.

Liang, H. (2011). *User Profiling based on Folksonomy Information in Web 2.0 for Personalised Recommender Systems*. Doctoral Dissertation, Queensland University of Technology, Brisbane.

Liang, H., Xu, Y., Li, Y., Nayak, R., & Weng, L. T. (2009a). Personalized recommender systems integrating social tags and item taxonomy. *Proceedings of the Joint Conference on Web Intelligence and Intelligent Agent Technology* (pp. 540–547).

Liang, H., Xu, Y., Li, Y., & Nayak, R. (2009b). Collaborative filtering recommender systems based on popular tags. *Proceedings of the 14th Australasian Document Computing Symposium*, Sydney.

Linden, G., Smith, B., & York, J. (2003). Amazon.com recommendations: Item-to-item collaborative filtering. *Internet Computing, IEEE, 7*(1), 76–80.

Liu, D.-R., & Shih, Y.-Y. (2005). Integrating AHP and data mining for product recommendation based on customer lifetime value. *Information and Management, 42*(3), 387–400.

Ma, S. (2008). *Improvements to personalized recommender systems*. Dissertation M.Phil, The University of Queensland, Australia.

Malone, T. W., Grant, K. R., Turbak, F. A., Brobst, S. A., & Cohen, M. D. (1987). Intelligent information-sharing systems. *Communications of the ACM, 30*(5), 390–402.

Manouselis, N., & Costopoulou, C. (2007). Experimental analysis of design choices in multi-attribute utility collaborative filtering. *International Journal of Pattern Recognition and Artificial Intelligence, 21*(2), 311–331.

Marinho, L. B., & Schmidt-Thieme, L. (2007). Collaborative tag recommendations. *Proceedings of the 31st Annual Conference of the Gesellschaft für Klassifikation* (pp. 533–540).

Marsh, S. (1994). *Formalising trust as a computational concept*. Dissertation Doctoral, University of Stirling, UK.

Massa, P., & Avesani, P. (2004). Trust-aware collaborative filtering for recommender systems. *Lecture Notes in Computer Science, 3290*, 492–508.

Massa, P., & Avesani, P. (2007). Trust-aware Recommender Systems. *Proceedings of the ACM Conference on Recommender Systems*, 17-24.

Massa, P., & Avesani, P. (2005). & Tiella, R. A Trust-enhanced Recommender System application: Moleskiing. Procedings of ACM Symposium on Applied Computing TRECK Track.

Massa, P., & Bhattacharjee, B. (2004). Using trust in recommender systems: An experimental analysis. *Proceedings of the 2nd International Conference on Trust Management*, Oxford, UK. 221-235.

McIlroy, T. (2010). The Information Explosion. Downloaded on 16 April, 2011 from http://www.thefutureofpublishing.com/images/uploadimages/Information_Explosion-08-15-10.pdf

McKnight, D. H., & Chervany, N. L. (2002). What trust means in e-commerce customer relationships: An interdisciplinary conceptual typology. *International Journal of Electronic Commerce, 6*(2), 35–59.

Middleton, S. E., Shadbolt, N. R., & Roure, D. C. D. (2004). Ontological user profiling in recommender systems. *ACM Transactions on Information Systems, 22*(1), 54–88.

Millen, D. R., Feinberg, J., & Kerr, B. (2006). *Dogear: Social bookmarking in the enterprise* (pp. 111–120). New York, NY: Proceedings of the SIGCHI Conference on Human Factors in Computing Systems.

Millen, D. R., Yang, M., Whittaker, S., & Feinberg, J. (2007). Social bookmarking and exploratory search. *Proceedings of the 10th European Conference on Computer Supported Cooperative Work*, 21-40.

Mira, K., & Dong-Sub, C. (2001). *Collaborative filtering with automatic rating for recommendation.* Pusan, Korea: Proceedings of the IEEE International Symposium on Industrial Electronics.

Mobasher, B., Dai, H., Luo, T., & Nakagawa, M. (2002). Discovery and Evaluation of Aggregate Usage Profiles for Web Personalization. *Data Mining and Knowledge Discovery, 6*(1), 61–82.

Montaner, M., Lopez, B., & Rosa, J. L. (2002). Opinion-based filtering through trust. *Proceedings of the 6$^{th}$ International Workshop on Cooperative Information Agents,* Madrid, Spain.

Montaner, M., Lopez, B., & Rosa, J. L. (2003). A taxonomy of recommender agents on the Internet. *Artificial Intelligence Review, 19*, 285–330.

Mui, L., Mohtashemi, M., & Halberstadt, A. (2002). A Computational model of trust and reputation. *Proceedings of the 35th Hawaii International Conference on System Science.*

Newcomb, T. M. (1961). *The Acquaintance Process.* Holt, Rinehart & Winston, New York.

Niwa, S., Doi, T., & Hon'Iden, S. (2006). Web page recommender system based on folksonomy mining. *Transactions of Information Processing Society of Japan, 47*(5), 1382–1392.

O'Donovan, J., & Smyth, B. (2005). Trust in recommender systems. *Proceedings of the 10th International Conference on Intelligent User Interfaces,*167-174.

Papagelis, M., Plexousakis, D., & Kutsuras, T. (2005). Alleviating the sparsity problem of collaborative filtering using trust inferences. *Proceedings of iTrust.* Springer. 224-239.

Park, S. T., Pennock, D., Good, N., & Decoste, D. (2006). Naïve filterbots for robust cold-start recommendations. *Proceedings of the 12$^{th}$ International Conference on Knowledge Discovery and Data Mining.*

Pazzani, M. J. (1999). A framework for collaborative, content-based and demographic filtering. *Artificial Intelligence Review, 13*(5–6), 393–408.

Pazzani, M. J., & Billsus, D. (1997). Learning and revising user profiles: The identification of interesting web sites. *Machine Learning, 27*(3), 313–331.

Pazzani, M. J., & Billsus, D. (2007). Content-based recommender systems. In P. Brusilovsky, A. Kobsa, & W. Nejdl (Eds.), *The Adaptive Web*. Berlin: Springer-Verlag.

Peel, E. (2009). *Online survey research about lesbian and bisexual women's experiences of pregnancy loss: Positive and pitfalls.* : British Psychological Society Division of Health Psychology Conference.

Peng, T., & Seng-cho, T. (2009). *iTrustU: A blog recommender system based on multi-faceted trust and collaborative filtering* (pp. 1278–1285). NY: Proceedings of the ACM Symposium on Applied Computing. New York.

Qu, L., Muller, C., & Gurevych, I. (2008). Using tag semantic network for keyphrase extraction in blogs. *Proceedings of the 17th ACM Conference on Information and Knowledge Management*, Napa Valley, CA, 1381-1382.

Rashid, A. M., Lam, S. K., & Karypis, G. (2006). *& Riedl, J.* ClustKNN: A highly scalable hybrid model- & memory-based CF algorithm. Proceedings of the KDD Workshop on Web Mining and Web Usage Analysis.

Rasmusson, L., & Janssen, S. (1996). Simulated social control for secure Internet commerce. Proceedings of the New Security Paradigms Workshop

Resnick, P., & Varian, H. R. (1997). Recommender Systems. *Communications of the ACM, 40,* 56–58.

Rich, E. (1979). User modeling via stereotypes. *Cognitive Science, 3*(4), 329–354.

Sabater, J., & Sierra, C. (2005). Review on computational trust and reputation models. *Artificial Intelligence Review, 24,* 33–60.

Salton, G. (Ed.). (1988). *Automatic text processing.* Boston: Addison-Wesley Longman Publishing Co., Inc.

Salton, G., & Buckley, C. (1988). Term weighting approaches in automatic text retrieval. *Information Processing and Management, 24*(5), 513–523.

Sarwar, B., Karypis, G., Konstan, J., & Reidl, J. (2000). Application of Dimensionality Reduction in Recommender Systems. *ACM Workshop on Web Mining for E-Commerce Challenges and Opportunities* August 2000, Boston, USA.

Sarwar, B., Karypis, G., Konstan, J., & Riedl, J. (2002). Recommender systems for large-scale e-commerce: Scalable neighborhood formation using clustering. *Proceedings of the 5th International Conference on Computer and Information Technology.*

Schafer, J. B., Konstan, J. A., & Riedl, J. (2001). E-Commerce Recommendation Applications. *Journal of Data Mining and Knowledge Discovery, 5*(1–2), 115–153.

Schickel-Zuber, V., & Faltings, B. (2005). *Heterogeneous attribute utility model: a new approach for modelling user profiles for recommendation systems.* Chicago, IL: Proceedings of the Workshop on Knowledge Discovery in the Web.

Schillo, M., Funk, P., & Rovatsos, M. (2000). Using trust for detecting deceitful agents in artificial societies. *Applied Artificial Intelligence* (Special Issue on Trust, Deception and Fraud in Agent Societies).

Schmitt, C., Dengler, D., & Bauer, M. (2002). The MAUT machine: an adaptive recommender system. Proceedings of the ABIS Workshop, Hannover, Germany

Selcuk, A., Uzun, E., & Pariente, M. R. (2004). *A reputation-based trust management system for P2P networks.* : IEEE International Symposium on Cluster Computing and the Grid.

Sen, S., Vig, J., & Riedl, J. (2009). Tagomenders: Connecting users to items through tags. *Proceedings of the 18th International World Wide Web Conference,* 671-680.

Shardanand, U., & Maes, P. (1995). Social information filtering: Algorithms for automating "word of mouth". *Proceedings of ACM CHI'95 Conference on Human Factors in. Computing Systems, 1,* 210–217.

Shepitsen, A., Gemmell, J., Mobasher, B., & Burke, R. D. (2008). *Personalized recommendation in social tagging systems using hierarchical clustering* (pp. 259–266). Lausanne, Switzerland: Proceedings of the ACM Conference on Recommender Systems.

Shepitsen, A., & Tomuro, N. (2009). Search in social tagging systems using ontological user profiles. *Proceedings of the 3rd International AAAI Conference on Weblogs and Social Media,* San Jose, California, USA.

Shmatikov, V., & Talcott, C. (2005). Reputation-based trust management. *Journal of Computer Security, 13*(1), 167–190.

Siersdorfer, S., & Sizov, S. (2009). Social recommender systems for web 2.0 folksonomies. *Proceedings of the 20th ACM Conference on Hypertext and Hypermedia,* Torino, Italy.

Sinha, R., & Sweringen, K. (2001). Comparing recommendations made by online systems and friends. *Proceedings of the DELOS-NSF Workshop on Personalization and Recommender Systems in Digital Libraries.* June 2001, Dublin, Ireland.

Snyder, C., & Fromkin, H. (1980). *Uniqueness: The Human Pursuit of Difference.* New York, NY, USA: Plenum.

Stolze, M., & Stroebel, M. (2003). Dealing with learning in e-Commerce product navigation and decision support: the teaching salesman problem. *Proceedings of the 2nd Interdisciplinary World Congress on Mass Customization and Personalization,* Munich, Germany.

Tavakolifard, M. (2010). Similarity-based Techniques for Trust Management. In Z. Usmani (Ed.), *Web Intelligence and Intelligent Agents* (pp. 233–250). Croatia: In-Tech.

Teacy, W. T. L. (2005). *An investigation into trust and reputation for agent-based virtual organisations*. Doctoral dissertation: University of Southampton, UK.

Terveen, L., Hill, W., Amento, B., McDonald, D., & Creter, J. (1997). PHOAKS: A system for sharing recommendations. *Communications of ACM, 40*(3), 59–62.

Tian, C.-Q., Zou, S.-H., Wang, W.-D., & Cheng, S.-D. (2008). Trust model based on reputation for peer-to-peer networks. *Journal on Communication, 29*(4), 63–70.

Towle, B., & Quinn, C. (2000). Knowledge Based Recommender Systems using Explicit User Models. *Proceedings of the AAAI Workshop on Knowledge-Based Electronic Markets, (AAAI Technical Report WS-00–04)*, Menlo Park, CA, 74-77.

Tso-Sutter, K. H. L., Marinho, L. B., & Schmidt-Thieme, L. (2008). *Tag-aware recommender systems by fusion of collaborative filtering algorithms* (pp. 1995–1999). USA: Proceedings of the ACM Symposium on Applied Computing.

Victor, P., Cornelis, C., De Cock, M., & Pinheiro da Silva, P. (2008). Gradual trust and distrust in recommender systems. Fuzzy Set and Systems, Vol 160. 1367-1382

Victor, P., De Cock, M., Cornelis, C., & Teredesai, A. (2008b). Getting cold start users connected in a recommender systems trust network. *Computational Intelligence in Decision and Control, 1*, 877–882.

Wang, Y., & Vassileva, J. (2007). A review on trust and reputation for web service selection. *Proceedings of the 27th International Conference on Distributed Computing Systems Workshops*.

Wei, Y. Z., Moreau, L., & Jennings, N. R. (2005). A market-based approach to recommender systems. *ACM Transactions on Information Systems, 23*(3), 227–266.

Weng, L. T. (2008). *Information Enrichment for Quality Recommender Systems*. Doctoral dissertation: Queensland University of Technology, Australia.

Weng, L. T., Xu, Y., Li, Y., & Nayak, R. (2008). Web information recommendation making based on item taxonomy. *Proceedings of the 10th International Conference on Enterprise Information Systems*, 20-28.

Wetzker, R., Umbrath, W., & Said, A. (2009). A hybrid approach to item recommendation in folksonomies. *Proceedings of the WSDM Workshop on Exploiting Semantic Annotations in Information Retrieval*, 25–29.

Wetzker, R., Zimmermann, C., Bauckhage, C., & Albayrak, S. (2010). I tag, you tag: Translating tags for advanced user models. Proceedings of the International Conference on Search and Data Mining

Wikipedia: Free encyclopedia built using Wiki software. Retrieved on 10 February 2011, from http://en.wikipedia.org/wiki/List_of_ social_networking_websites.

Xiong, L. (2005). *Resilient Reputation and Trust Management: Models and Techniques*. Georgia Institute of Technology, USA: Doctoral dissertation.

Xu, D., Zhang, L., & Luo, J. (2010). Understanding multimedia content using web scale social media data. *Proceedings of the 18th International Conference on Multimedia*, Firenze, Italy, 1777-1778

Xue, W., & Fan, Z. (2008). A new trust model based on social characteristic and reputation mechanism for the semantic web. Proceedings of the Workshop on Knowledge Discovery and Data Mining

Yeung, C. M. A., Gibbins, N., & Shadbolt, N. (2008). A study of user profile generation from folksonomies. Proceedings of the Social Web and Knowledge Management Workshop at World Wide Web Conference

Yu, B., & Singh, M. P. (2002). Distributed reputation management for electronic commerce. *Computational Intelligence, 18*(4), 535–549.

Zhang, Z., Zhou, T., & Zhang, Y. (2010). Personalized recommendation via integrated diffusion on user-item-tag tripartite graphs. *Physica A 389*, Elsevier, 179-186.

Zhen, Y., Li, W., & Yeung, D. (2009). TagiCoFi: Tag Informed Collaborative Filtering. *Proceedings of the third ACM Conference on Recommender Systems, 69-76*.

Ziegler, C. N., & Golbeck, J. (2007). Investigating interactions of trust and interest similarity. *Decision Support Systems, 43*, 460–475.

Ziegler, C. N. (2005). *Towards Decentralized Recommender Systems*. Doctoral dissertation: University of Freiburg, Germany.

Ziegler, C. N., & Lausen, G. (2004). Analyzing correlation between trust and user similarity in online communities. *iTrust 2004*, 251-265.

Zigoris, P., & Zhang, Y. (2006). Bayesian adaptive user profiling with explicit & implicit feedback. *Proceedings of ACM 15th Conference on Information and Knowledge Management*. Arlington, Virginia, USA.